Underwater Work

Edited by Sérgio António Neves Lousada

Published in London, United Kingdom

IntechOpen

Supporting open minds since 2005

Underwater Work
http://dx.doi.org/10.5772/intechopen.83282
Edited by Sérgio António Neves Lousada

Assistant to the Editor: Rafael Camacho

Contributors
Mario Alberto Alberto Jordán, Shaik Asif Hossain, Monir Hossen, Kimon Papadimitriou, Elpida Karadimou, Alexandros Tourtas, Ralph Schill, Sérgio António Neves Lousada, Rafael Camacho, Josué Suárez Palacios

Notice
Statements and opinions expressed in the chapters are these of the individual contributors and not necessarily those of the editors or publisher. No responsibility is accepted for the accuracy of information contained in the published chapters. The publisher assumes no responsibility for any damage or injury to persons or property arising out of the use of any materials, instructions, methods or ideas contained in the book.

First published in London, United Kingdom, 2021 by IntechOpen
IntechOpen is the global imprint of INTECHOPEN LIMITED, registered in England and Wales, registration number: 11086078, 5 Princes Gate Court, London, SW7 2QJ, United Kingdom
Printed in Croatia

British Library Cataloguing-in-Publication Data
A catalogue record for this book is available from the British Library

Additional hard and PDF copies can be obtained from orders@intechopen.com

Underwater Work
Edited by Sérgio António Neves Lousada
p. cm.
Print ISBN 978-1-78985-222-6
Online ISBN 978-1-78985-229-5
eBook (PDF) ISBN 978-1-78985-230-1

Meet the editor

Sérgio António Neves Lousada has an international Ph.D. in Civil Engineering (Hydraulics). He teaches Hydraulics, Environment, and Water Resources and Construction at the University of Madeira, Portugal. He has published articles and books and participated in events mainly in the areas of hydraulics, urban planning, and land management. Furthermore, he collaborates with the Environmental Resources Analysis Research Group (ARAM), University of Extremadura (UEx); VALORIZA - Research Center for the Enhancement of Endogenous Resources, Polytechnic Institute of Portalegre (IPP), Portugal; CITUR - Madeira - Centre for Tourism Research, Development and Innovation, Madeira, Portugal; and Institute of Research on Territorial Governance and Inter-Organizational Cooperation, Dąbrowa Górnicza, Poland. Moreover, he holds an International master's degree in Ports and Coasts Engineering.

Contents

Preface

The demand for underwater work is increasing. As such, this book provides guidelines for developing a successful underwater work curriculum, designing an innovative learning and teaching method, and promoting consistent standards in underwater work education. It includes case studies of underwater work, integration of blended e-learning, and sustainable and social innovation in underwater work learning experiences. The five chapters focus on cutting-edge research and provide the reader with a broad overview of the current state of development in underwater work design and methods. The chapters in this volume include the relevant technical, sustainable, and social innovations that have a significant influence on society and the stakeholders. This edited book consists of 5 chapters focusing on underwater work. A brief summary of each part is given below.

Chapter 1 "Introductory Chapter: Underwater Ordeals" is a brief introduction defining underwater work, its trades and crafts, and associated challenges.

Chapter 2 "Cross-Correlation-Based Fisheries Stock Assessment Technique: Utilization of Standard Deviation of Cross-Correlation Function as Estimation Parameter with Four Acoustic Sensors" presents the cross-correlation-based fisheries stock assessment technique, which uses the mean and ratio of the standard deviation to the mean of cross-correlation function (CCF) as the estimation parameter. However, this chapter utilizes only the standard deviation of CCF as a parameter to estimate population size. The authors use four acoustic sensors and a chirping sound commonly generated by damselfish (Dascyllus aruanus), humpback whales (Megaptera novaeangliae), dugongs (Dugong dugon), and other species to accomplish the simulations. Results show that a robust estimation can be obtained using the standard deviation of CCF as an estimation parameter even when the distances between acoustic sensors are small.

Chapter 3 "Diving as a Scientist: Training, Recognition, Occupation - The 'Science Diver' Project" describes the challenge of conducting scientific work under water. From collecting samples to protecting underwater cultural heritage sites, scientific divers need to address issues concerning scientific methodology, diving safety, professional acknowledgment, training, and legal implications. All these matters are handled in different ways depending on factors like region, organizations involved, legal framework, diving philosophy, and so on, producing a diverse framework on scientific diving as a distinct type of underwater work. The Science DIVER project's main objective is to study and analyze this fragmented landscape to provide insight and suggestions for a commonly accepted framework that will promote scientific diving as a means of forwarding knowledge both within the scientific community and its interaction with the public.

Chapter 4 "Progressive Underwater Exploration with a Corridor-Based Navigation System" focuses on the exploration of underwater environments by means of autonomous submarines like autonomous underwater vehicles (AUVs) using vision-based navigation. An approach called Corridor SLAM (C-SLAM) was developed for this purpose. It implements a global exploration strategy that consists of first creating a trunk corridor on the seabed and then branching as far as possible in different directions to expand the explored region. The system guarantees the safe return of the vehicle to the starting point by taking into account a metric of the corridor lengths that are related to their energy autonomy. Experimental trials in a basin with underwater scenarios demonstrate the feasibility of the approach.

Chapter 5 "Underwater Technical Inspections Using ROV Applied to Maritime and Coastal Engineering: The Study Case of Canary Islands" describes how underwater technical inspections using remotely operated vehicles (ROV) have an important role in the design, construction, maintenance, and repair of maritime and coastal infrastructures, trough video recording, digital photographs, collection of technical data, and underwater topographic survey providing support for consultancy studies and projects and technical advice and appraisals. Routine inspections are key to the maintenance of any submerged infrastructure. The importance of this type of inspection is increasing every day, but divers are also placed in increasingly dangerous scenarios to carry out this type of work. Inspections of underwater structures (as in dams, bridges, reservoirs, breakwaters, piers, oil rigs, etc.) have always been arduous and difficult, and often dangerous, but today underwater drones offer solutions that eliminate the risk faced by divers as well as greatly reduce the high costs involved in such inspections.

We hope this volume will enhance readers' understanding and practice of underwater work and processes.

Sérgio António Neves Lousada and Rafael Freitas Camacho
UMa - Faculdade de Ciências Exatas e da Engenharia,
Campus Universitário da Penteada,
Funchal, Portugal

Introductory Chapter: Underwater Ordeals

Sérgio Lousada and Rafael Camacho

1. Underwater works

Underwater work is work done underwater, generally by divers during diving operations, but includes work done underwater by remotely operated vehicles and manned submersibles.

The versatility and multifarious skills of underwater works means that it is possible to operate over a wide range of activities, working in hyperbaric conditions or in confined spaces. The divers' experience in the field and their detailed knowledge of diving procedures enables them to operate in highly specific segments [1]:

- inspection of civil-engineering structures;

- undersea foundations and welds;

- ship hull inspections and raising of wrecks;

- work in hostile and nuclear environments;

- dam inspections using an ROV (Remotely Operated Vehicle);

- installing or commissioning outfalls, undersea conduits, and cables.

2. Trades and crafts

Virtually, all the civil-engineering trades and crafts can be transposed to underwater work, as in the case of high-pressure cleaning, cementing, welding, cutting, among others [1]. Therefore, all professional diving occupations have a few skills commonly used:

- underwater navigation;

- underwater searches;

- rigging and lifting;

- inspection, measuring, and recording;

- and the use of basic hand tools.

Some skills are specific to specialist occupations such as: erecting formwork and shuttering (civils), oxy-arc cutting (salvage, ships husbandry, offshore), hydraulic bolt-tensioning (offshore oil and gas), bomb disposal (military, public safety), search and rescue (public safety, police), and site surveys and mapping (scientific, archeology).

3. Ordeals

Most construction projects involving professional divers are engineered by road, canal, and port engineers, but only a few know in depth the risks inherent in the underwater work performed by professional divers [2].

As in any profession, engineers need a permanent updating in their area of expertise through continuous training [2].

One of these subareas is underwater engineering. However, for several years, engineering projects have suffered from a lack of rigor in their approach to underwater work, both at budget level, constructive procedures and in terms of safety and health [2].

It is believed that this is so, mainly due to the ignorance of the exceptional conditions that the hyperbaric environment imposes throughout the activity and the legal framework that regulates it. This means that the tasks are tendered with significant shortcomings that hinder their subsequent execution in adequate conditions of safety and economic viability [2].

The main objective of this book is precisely conveying the on-going constructive procedures, methods and methodologies, the equipment, the limitations, and the specificities that the hyperbaric environment has, where the diver develops his work, that so much condition an underwater work.

Author details

Sérgio Lousada[1,2,3,4*] and Rafael Camacho[1,5]

1 Faculty of Exact Sciences and Engineering (FCEE), Department of Civil Engineering and Geology (DECG), University of Madeira (UMa), Funchal, Portugal

2 VALORIZA - Research Centre for Endogenous Resource Valorization, Portalegre, Portugal

3 Institute of Research on Territorial Governance and Inter-Organizational Cooperation, Dąbrowa Górnicza, Poland

4 CITUR - Madeira - Centre for Tourism Research, Development and Innovation, Madeira, Portugal

5 IHM - Investimentos Habitacionais da Madeira, EPERAM, Portugal

*Address all correspondence to: slousada@staff.uma.pt

IntechOpen

References

[1] VINCI Construction. Expertises, Know-How, Underwater Works [Internet]. 2019. Available from: https://www.vinci-construction-maritime-fluvial.fr/en/categorie/65/10/know-how/underwater-works.html [Accessed: 20 July 2019]

[2] Colegio de Ingenieros de Caminos, Canales y Puertos. Jornada Técnica Sobre Buceo Profesional en la Ingeniería [Internet]. 2019. Available from: http://www3.ciccp.es/wp-content/uploads/2017/06/JORNADA_BUCEO_INGENIERIA.pdf [Accessed: 20 July 2019]

Chapter 2

Cross-Correlation-Based Fisheries Stock Assessment Technique: Utilization of Standard Deviation of Cross-Correlation Function as Estimation Parameter with Four Acoustic Sensors

Shaik Asif Hossain and Monir Hossen

Abstract

In the past, cross-correlation-based fisheries stock assessment technique utilized the mean and the ratio of standard deviation to the mean of cross-correlation function (CCF) as estimation parameter. However, in this paper, we have utilized only standard deviation of CCF as estimation parameter to estimate the population size. We utilized four acoustic sensors and considered chirp sound which is commonly generated by damselfish (*Dascyllus aruanus*), humpback whales (*Megaptera novae-angliae*), dugongs (*Dugong dugon*), etc., species to accomplish the simulations. We found that a robust estimation can be obtained using standard deviation of CCF as estimation parameter even when the distances between acoustic sensors are small.

Keywords: acoustic sensor, bins, chirp, fisheries stock assessment, standard deviation

1. Introduction

Passive acoustic monitoring of fish abundance is an emerging field of research among the conservation researchers and marine ecologists. It has upgraded understanding of the temporal distribution and repertoire of soniferous fish and mammals [1–2]. Generally, passive acoustic monitoring is used to have an insight about the population size of soniferous fish and mammals, which are problematic to locate using visual sampling techniques [3–6] in a certain marine area. These types of fishery surveys utilize the advantage of sound production nature of many species of fish and mammals which possess natural acoustic tags. It has the merit of being a non-destructive and non-invasive monitoring technique, unlike the conventional fisheries stock assessment methods, that is, mark recapture techniques, environmental DNA, visual census, echo, minnow traps, etc. [7–8]. Generally, mechanical instrument-based conventional fishery surveys suffer from poor accuracy, time consuming nature, overly human interaction, costly instruments, etc., which can be overcome by passive acoustic monitoring techniques. Passive monitoring can

Hybrid Computer

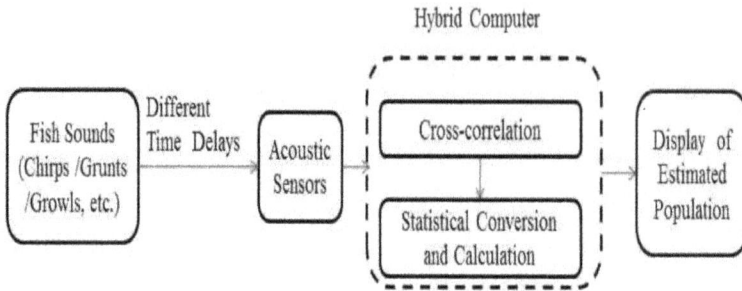

Figure 1.
Simplified block diagram of cross-correlation fisheries-based stock assessment system.

provide unbiased data on the location and movement of sound producing source in underwater situations [9]. Low-frequency (<10 kHz) acoustic sensors, that is, hydrophones, are used to detect natural sound production by fish and mammals [10]. Usually, fish sound is associated with courtship, feeding or aggressive encounters [10]. Researchers categorized the sound types of fish and mammals by different names, that is, chirps, pops, grunts, whistles, growls, hoots, etc., which are associated with their frequency and temporal characteristics [11].

However, cross-correlation-based fisheries stock assessment technique, a passive acoustic survey technique, was proposed in [12–16]. In this technique, the sound signals of vocalizing fish and mammals are processed to estimate their population size [17]. This statistics-based technique has the potential to resolve some main drawbacks of conventional techniques like complexity, reliance on human interaction, time consuming nature of estimation, sensitivity, high cost, etc. A simplified block diagram representation of this technique is illustrated in **Figure 1**.

In the past, the researchers associated with this technique utilized the mean of CCF [12] ratio of standard deviation to the mean of CCF [13–20] to estimate population size. In this paper, we have introduced standard deviation of CCF as estimation parameter to perform our desired estimation. We considered four acoustic sensors case [21], that is, hydrophones, in this research. For four acoustic sensors case, different types of topologies, that is, acoustic sensors in line, acoustic sensors in a rectangular shape, acoustic sensors in a triangular shape, are possible. Similarly, Acoustic sensors in a triangular shape can be a square shape, a rhombus shape or a trapezoidal shape. In this paper, we considered acoustic sensors in line case (ASL case). The main reason of considering four acoustic sensors is increasing number of cross-correlation function ensures better accuracy in this technique [14]. Likewise, from diverse sound types of fish and mammals, we considered chirp sound which is commonly generated by damselfish (*Dascyllus aruanus*), humpback whales (*Megaptera novaeangliae*), dugongs (*Dugong dugon*), etc., species [11]. We organize this paper as firstly, to state the theoretical procedure of our proposed methodology and finally, the theory will be evaluated by simulation. We used MATLAB simulation environment to accomplish our simulation in this study.

2. Utilization of the CCF

The formulation of cross-correlation of sound signals of fish and mammals is analogous to the formulation of cross-correlation of Gaussian signal [22], which are the starting materials to estimate the population size. Chirp sound of fish and mammals are received by the acoustic sensor and recorded in the associated computer in which cross-correlation is executed. Transmission and reception of sound signals are performed for a time frame, called "signal length." Sound (chirp) generating fish and mammals are considered as the sources of sound signals and N fish and mammals are distributed over the volume of a large sphere, the center of which lies halfway between the acoustic sensors. A typical scenario of fish and mammals distribution is shown in **Figure 2**.

In the water medium, a constant propagation speed S_p of sound is considered [23]. **Figure 3** shows an example of 3D estimation area under water space with a single fish N_1 and four acoustic sensors H_1, H_2, H_3, and H_4. We considered the coordinates of H_1, H_2, H_3, and H_4 are (x_1, y_1, z_1), (x_2, y_2, z_2), (x_3, y_3, z_3), and (x_4, y_4, z_4) respectively, whereas the coordinate of the fish is (a, b, c). The distance between the acoustic sensors can be calculated as follows:

$$d_{DBS12} = \sqrt{\left(x_1 - x_2\right)^2 + \left(y_1 - y_2\right)^2 + \left(z_1 - z_2\right)^2} \tag{1}$$

$$d_{DBS23} = \sqrt{\left(x_2 - x_3\right)^2 + \left(y_2 - y_3\right)^2 + \left(z_2 - z_3\right)^2} \tag{2}$$

$$d_{DBS34} = \sqrt{\left(x_3 - x_4\right)^2 + \left(y_4 - y_4\right)^2 + \left(z_3 - z_4\right)^2} \tag{3}$$

Here, d_{DBS12} = distance between H_1 and H_2, d_{DBS23} = distance between H_2 and H_3, and d_{DBS34} = distance between H_3 and H_4.

Let us consider, a sound signal coming from N_1 is $S_1(t)$, which is finite in length. The signal received by acoustic sensors H_1, H_2, H_3, and H_4 are Sr_{11}, Sr_{12}, Sr_{13}, and Sr_{14}, respectively:

$$S_{r_{11}}(t) = \alpha_{11} S_{11}\left(t - \tau_{11}\right), \tag{4}$$

$$S_{r_{12}}(t) = \alpha_{12} S_{12}\left(t - \tau_{12}\right), \tag{5}$$

$$S_{r_{13}}(t) = \alpha_{13} S_{13}\left(t - \tau_{13}\right), \tag{6}$$

$$S_{r_{14}}(t) = \alpha_{14} S_{14}\left(t - \tau_{14}\right), \tag{7}$$

where α_{11}, α_{12}, α_{13}, and α_{14} are the attenuation due to absorption and dispersion in the medium, and τ_{11}, τ_{12}, τ_{13}, and τ_{14} are the respective time delays for the acoustic signals to reach the acoustic sensors. For four acoustic sensors ASL case, the cross-correlation among the acoustic sensors is taken place for three times, i.e., between sensors H_1 and H_2, H_2 and H_3, and H_3 and H_4. So, the total number of CCF is three.

Therefore, the CCFs are:

$$C_1(\tau) = \int_{-\infty}^{+\infty} S_{11}(t) S_{12}\left(t - \tau_{11}\right) d\tau \tag{8}$$

$$C_2(\tau) = \int_{-\infty}^{+\infty} S_{12}(t) S_{13}\left(t - \tau_{12}\right) d\tau \tag{9}$$

$$C_3(\tau) = \int_{-\infty}^{+\infty} S_{13}(t) S_{14}\left(t - \tau_{13}\right) d\tau \tag{10}$$

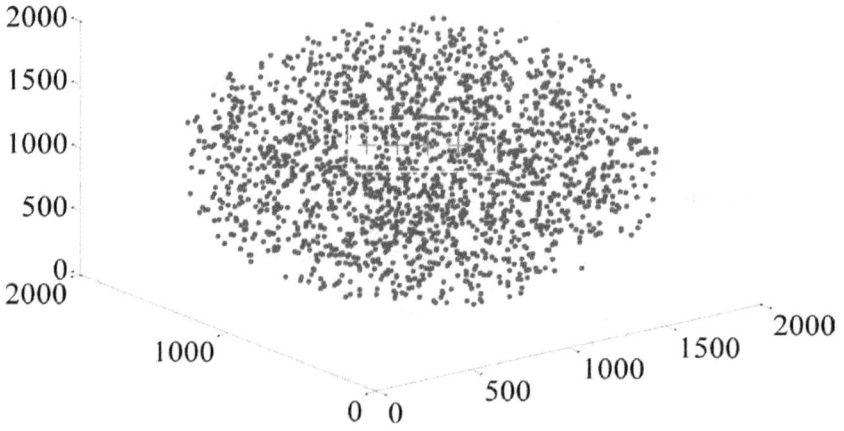

Figure 2.
Distribution of fish and mammals with four acoustic sensors, that is, four pluses (++++).

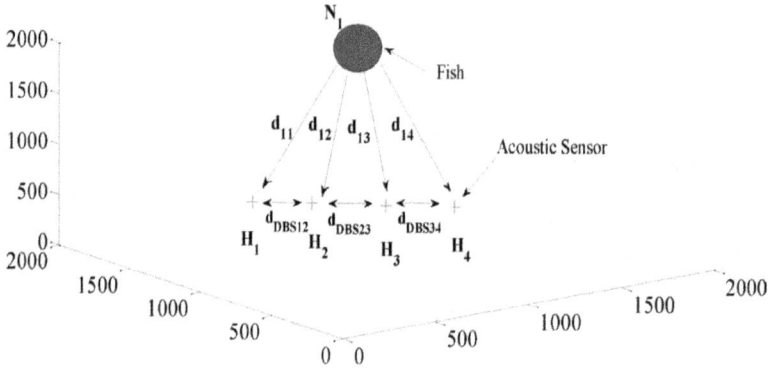

Figure 3.
Diagram of a single fish with four acoustic sensors, H_1, H_2, H_3, and H_4.

To find out the CCFs for N number of fish and mammals, we have to take the total sound signals received by the four acoustic sensors.

Thus, the composite signals received by H_1, H_2, H_3, and H_4 are:

$$S_{rt1} = \sum_{j=1}^{N} \alpha_{j1} S_j \left(t - \tau_{j1} \right) \tag{11}$$

$$S_{rt2} = \sum_{j=1}^{N} \alpha_{j2} S_j \left(t - \tau_{j2} \right) \tag{12}$$

$$S_{rt3} = \sum_{j=1}^{N} \alpha_{j3} S_j \left(t - \tau_{j3} \right) \tag{13}$$

$$S_{rt4} = \sum_{j=1}^{N} \alpha_{j4} S_j \left(t - \tau_{j4} \right) \tag{14}$$

Therefore, the total CCFs are:

$$C_{12} (\tau) = \int_{-\infty}^{+\infty} S_{rt1} (t) S_{rt2} (t - \tau) d\tau \tag{15}$$

$$C_{23}(\tau) = \int_{-\infty}^{+\infty} S_{rt2}(t) S_{rt3}(t-\tau) d\tau \qquad (16)$$

$$C_{34}(\tau) = \int_{-\infty}^{+\infty} S_{rt3}(t) S_{rt4}(t-\tau) d\tau \qquad (17)$$

This is the form of series of delta functions because in cross-correlation procedure one sound signal is the delayed copy of another [22].

3. Theoretical estimation from standard deviation of CCF

As we considered chirp generating fish and mammals to estimate their population size, an introduction to chirp signal is an important task in this perspective. Chirps belong to a swept-frequency sound signal, which possess a time varying frequency. From a sound analysis of *Plectroglyphidodon lacrymatus* and *Dascyllus aruanus* species of damselfish, It was seen that the produced chirps by them was consisted of trains of 12–42 short pulses of 3–6 cycles [12, 24]. The durations varied from 0.6 to 1.27 ms where the peak frequency varied from 3400 to 4100 Hz [25]. Such a sound signal can be represented as [8, 12, 13]:

$$X(t) = A\cos\left[2\pi\left\{\frac{(f_2-f_1)t^2}{2d} + f_1 t\right\} + P\right] \qquad (18)$$

where f_1 = starting frequency in Hz, f_2 = ending frequency in Hz, d = duration in second, P = starting phase, and A = amplitude.

However, the mean of CCF can be expressed by ensemble average of the chirp-signal cross-correlation as [22].

$$\langle C(t) \rangle = Q_T T_r v \int_{-\infty}^{+\infty} d\vec{r}_s \times \delta\left(t + \frac{|\vec{r}_a - \vec{r}_s|}{S_p} - \frac{|\vec{r}_b - \vec{r}_s|}{S_p}\right), \qquad (19)$$

where Q_T represents the acoustic power of the received signals from the sources taken to be constant over time and space, v is the creation rate of the sources whose unit is unit time per unit volume, T_r is the total recording time, $\vec{r}_s$ is the path length of sources from the origin, $\vec{r}_a$ is the path length of first acoustic sensor from the origin, and $\vec{r}_b$ is the path length of second sensor from the origin.

Now, the variance of the CCF can be defined as [22]:

$$Var(C(t)) = \langle C^2(t) \rangle - \langle C(t) \rangle^2, \qquad (20)$$

where $\langle C(t) \rangle^2$ and $\langle C^2(t) \rangle$ are defined in Eqs. (21) and (22), respectively, as [22]:

$$\langle C(t) \rangle^2 = Q_T^2 v^2 \left(\int_{-T_r/2}^{+T_r/2} dt \int_{-\infty}^{+\infty} d\vec{r}_1 \int_{-\infty}^{+t} d\tau_1 G(\vec{r}_1, \vec{r}_a; t-\tau_1) \times G(\vec{r}_1, \vec{r}_b; \tau + t - \tau_1) \right) \times$$
$$\left(\int_{-T_r/2}^{+T_r/2} d\tilde{t} \int_{-\infty}^{+\infty} d\vec{r}_3 \int_{-\infty}^{+t} d\tau_3 G(\vec{r}_3, \vec{r}_a; t-\tau_3) \times G(\vec{r}_3, \vec{r}_b; \tau + \tilde{t} - \tau_3) \right) \qquad (21)$$

and

$$
\langle C^2(t) \rangle = Q_T v^2 \left(\int\limits_{-T_r/2}^{+T_r/2} dt \int\limits_{-\infty}^{+\infty} d\vec{r}_1 \int\limits_{-\infty}^{+t} d\tau_1 G(\vec{r}_1, \vec{r}_a; t - \tau_1) \times G(\vec{r}_1, \vec{r}_b; \tau + t - \tau_1) \right) \times
$$

$$
\left(\int\limits_{-T_r/2}^{+T_r/2} d\tilde{t} \int\limits_{-\infty}^{+\infty} d\vec{r}_3 \int\limits_{-\infty}^{+t} d\tau_3 G(\vec{r}_3, \vec{r}_a; t - \tau_3) \times G(\vec{r}_3, \vec{r}_b; \tau + \tilde{t} - \tau_3) \right) +
$$

$$
Q_T v^2 \left(\int\limits_{-T_r/2}^{+T_r/2} dt \int\limits_{-\infty}^{+\infty} d\tilde{t} \left(\int\limits_{-\infty}^{+\infty} d\vec{r}_1 \int\limits_{-\infty}^{+t} d\tau_1 G(\vec{r}_1, \vec{r}_a; t - \tau_1) \times G(\vec{r}_1, \vec{r}_a; \tilde{t} - \tau_1) \right) \right) \times \qquad (22)
$$

$$
\left(\int\limits_{-\infty}^{+\infty} d\vec{r}_1 \int\limits_{-\infty}^{+t} d\tau_1 G(\vec{r}_1, \vec{r}_b; t - \tau_1) \times G(\vec{r}_1, \vec{r}_a; \tilde{t} - \tau_1) \right) +
$$

$$
Q_T v^2 \left(\int\limits_{-T_r/2}^{+T_r/2} dt \int\limits_{-\infty}^{+\infty} d\tilde{t} \left(\int\limits_{-\infty}^{+\infty} d\vec{r}_1 \int\limits_{-\infty}^{+t} d\tau_1 G(\vec{r}_1, \vec{r}_a; t - \tau_1) \times G(\vec{r}_1, \vec{r}_b; \tilde{t} - \tau_1) \right) \right) \times
$$

$$
\left(\int\limits_{-\infty}^{+\infty} d\vec{r}_2 \int\limits_{-\infty}^{+t} d\tau_2 G(\vec{r}_2, \vec{r}_b; t - \tau_2) \times G(\vec{r}_2, \vec{r}_b; \tilde{t} - \tau_2) \right),
$$

where $G(.)$ = Green's function. The other parameters signify their usual meanings [22].

Therefore, we can get the standard deviation, σ of the CCF as we know that standard deviation is the square root of the variance.

$$
\sigma = \sqrt{Var(C(t))} = \sqrt{\langle C^2(t) \rangle - \langle C(t) \rangle^2} \qquad (23)
$$

However, to analyze the random signal cross-correlation problem to find the standard deviation in the above way is very hard. Therefore, the problem can be reframed as a binomial probability problem which can make the analysis simpler. Since, cross-correlation function follows the binomial probability distribution in which the parameters are the number of balls, that is, fish and mammals, N, and one on the number of bins, b; therefore, the standard deviation, σ of the CCF is defined as bellow [22]:

$$
\sigma = \sqrt{N \times \frac{1}{b} \times \left(1 - \frac{1}{b}\right)}, \qquad (24)
$$

where N is the number of fish and mammals and b is the number of bins. Here, b can be achieved from the following Eq. [22]:

$$
b = \frac{2 \times d_{DBS} \times S_R}{S_P} - 1, \qquad (25)
$$

where S_R is the sampling rate, d_{DBS} is the distance between equidistant sensors, and S_P is the speed of sound propagation.

From Eq. (25), we can write the following formula:

$$
N = \frac{b^2 \times \sigma^2}{b - 1} \qquad (26)
$$

Therefore, if σ is available from simulation, the estimated population size of fish and mammals, N will be found from Eq. (26).

Now, for four acoustic sensors ASL case, the final standard deviation will be found from the average of σ_1, σ_2, and σ_3.

$$\sigma_{Average}^{3CCF} = \frac{\sigma_1 + \sigma_2 + \sigma_3}{3} \tag{27}$$

Thus, from Eq. (26), we can obtain that

$$N = \frac{b^2 \times \left(\sigma_{Average}^{3CCF}\right)^2}{b-1} \tag{28}$$

Therefore, if σ is available from simulation, N will be found from Eq. (28).

4. Simulation and discussion

Simulations were executed considering that four acoustic sensors lay on the center of a sphere. We also considered a uniform random distribution of fish and mammals. Thousand iterations were averaged to accomplish the simulated results. To ease the simulation, the power difference among the acoustic pulses transmitted by each fish and mammal was considered negligible. Here, we considered $d_{DBS12} = d_{DBS23} = d_{DBS34} = d_{DBS}$. The parameters used in MATLAB simulation are introduced in **Table 1**.

Figure 4 shows the theoretical and corresponding simulated results for the population estimation of fish and mammals in terms of the estimation parameter σ of CCF. The solid lines designate the theoretical results, and the stars, circles, squares, and triangles correspond the simulated results. The variations of b are as results of varying d_{DBS} in the four different **Figures 4(a)–(d)**. The other parameters are same for all the figures.

Figure 5 shows the difference between theoretical and simulated population size of fish and mammals for $b = 79$. In this figure, the solid line indicates the theoretical results, and the triangles are corresponding to the simulated results. From **Figure 5**, it can be seen that the theoretical and simulated results are closely stayed to each other, which signifies the strength of this population estimation method. Similarly, we can see that the number of bins, b has an impact on the estimation parameter, which is obvious from Eq. (28). We can see that the value of the standard deviation

Parameters	Value
Dimension of the sphere	2000 m
d_{DBS}	0.25, 0.5, 0.75, 1 m
S_P	1500 m/s
S_R	60 kSa/s
Absorption coefficient, a	1 dBm^{-1}
dispersion factor, k	0
b	19, 39, 59, 79

Table 1.
Parameters used in Matlab simulation.

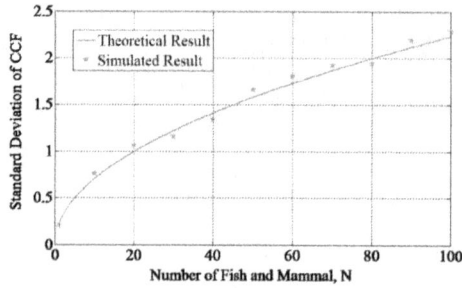

(a)

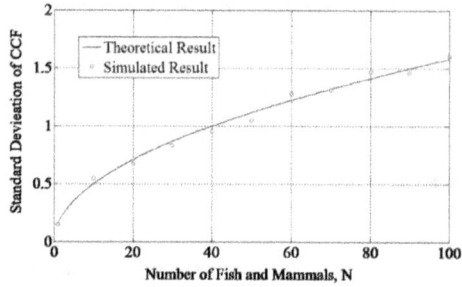

(b)

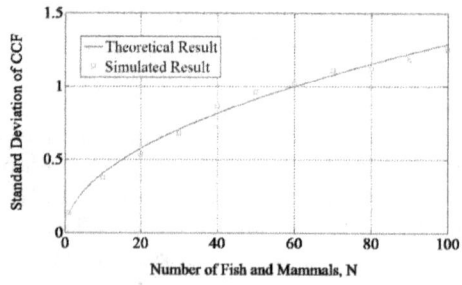

(c)

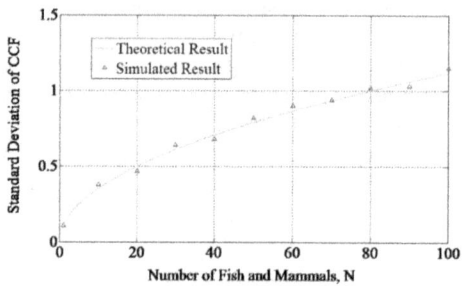

(d)

Figure 4.
Number of fish and mammals vs. σ of CCF (a) b = 19 (d$_{DBS}$ = 0.25 m and S$_R$ = 60 kSa/s) (b) b = 39 (d$_{DBS}$ = 0.5 m and S$_R$ = 60 kSa/s), (c) b = 59 (d$_{DBS}$ = 0.75 m and S$_R$ = 60 kSa/s), and (d) b = 79 (d$_{DBS}$ = 1 m and S$_R$ = 60 kSa/s).

is lower in case of higher *b* and vice-versa and the simulated results are closer with the theoretical lines also. The figures also illustrate that a very short distance, even to place a single fish between them, between the acoustic sensors can also give a good estimation using this technique.

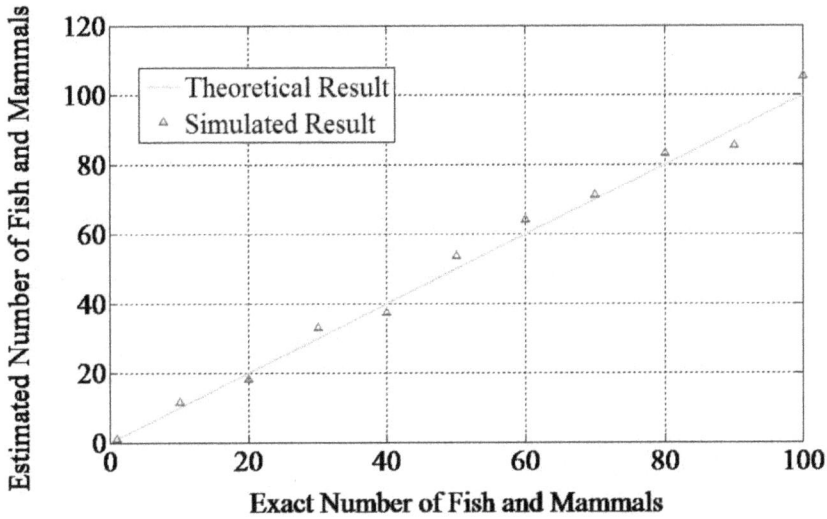

Figure 5.
Exact number of fish and mammals vs. estimated number of fish and mammals for b = 79 (d_{DBS} = 1 m and S_R = 60 kSa/s).

However, our work has some limitations, for example, assuming the delays to be integer, negligence of multipath interference, consideration of negligible amount of power difference among the fish sound pulses during transmitting time.

5. Conclusion

Passive acoustic monitoring is a potential tool to survey the population size of fish and mammals in a certain marine area. It can overcome the major drawbacks of conventional techniques. Cross-correlation-based stock assessment technique is also a passive acoustic survey technique dedicated to fish and mammals. An investigation on this technique with different estimation parameters was the cardinal goal of this research. To do that, we performed our desired estimation with standard deviation of CCF as estimation parameter. The small difference between theoretical and simulated results proved that it is highly possible to pursue this passive monitoring technique utilizing standard deviation of CCF as estimation parameter. Here, we considered four acoustic sensors because from the previous research, we found that an increasing number of CCF ensures better accuracy using this technique. In this paper, we considered four different numbers of bins to show its impact on estimation also. It is shown that a robust estimation is possible using standard deviation of CCF as estimation parameter even when the distances between acoustic sensors are small. Therefore, during practical implementation of this technique, these findings will contribute significantly.

Author details

Shaik Asif Hossain* and Monir Hossen
Department of Electronics and Communication Engineering, Khulna University of
Engineering and Technology, Khulna, Bangladesh

*Address all correspondence to: asifruete@gmail.com

IntechOpen

References

[1] Riera A, Pilkington JF, Ford JK, Stredulinsky EH, Chapman NR. Passive acoustic monitoring off Vancouver Island reveals extensive use by at-risk resident killer whale (*Orcinus orca*) populations. Endangered Species Research. 2019;**39**:221-234

[2] Palmer KJ, Brookes KL, Davies IM, Edwards E, Rendell L. Habitat use of a coastal delphinid population investigated using passive acoustic monitoring. Aquatic Conservation: Marine and Freshwater Ecosystems. 2019;**29**:254-270

[3] Gebbie J, Siderius M, Allen JS III. A two-hydrophone range and bearing localization algorithm with performance analysis. The Journal of the Acoustical Society of America. 2015;**137**(3):1586-1597

[4] Mouy X, Cabrera De Leo F, Juanes F, Dosso SE. Acoustic estimation of the biodiversity of fish and invertebrates. The Journal of the Acoustical Society of America. 2018;**144**(3):1921-1921

[5] Mouy X, Rountree R, Juanes F, Dosso SE. Cataloging fish sounds in the wild using combined acoustic and video recordings. The Journal of the Acoustical Society of America. 2018;**143**(5):EL333–EL339

[6] Mouy X, Rountree RA, Juanes F, Dosso SE. Passive acoustic localization of fish using a compact hydrophone array. The Journal of the Acoustical Society of America. 2017;**141**(5):3863-3863

[7] Putland RL, Mackiewicz AG, Mensinger AF. Localizing individual soniferous fish using passive acoustic monitoring. Ecological Informatics. 2018;**48**:60-68

[8] Hossain SA, Hossen M. A technical review on fish population estimation techniques: Non acoustic and acoustic approaches. Akustika. 2019;**31**:87-103

[9] Locascio JV, Mann DA. Localization and source level estimates of black drum (*Pogonias cromis*) calls. The Journal of the Acoustical Society of America. 2011;**130**(4):1868-1879

[10] Luczkovich JJ, Mann DA, Rountree RA. Passive acoustics as a tool in fisheries science. Transactions of the American Fisheries Society. 2008;**137**(2):533-541

[11] Amorim MCP. Diversity of sound production in fish. Communication in Fishes. 2006;**1**:71-104

[12] Hossain SA, Hossen M, Anower S. Estimation of damselfish biomass using an acoustic signal processing technique. Journal of Ocean Technology. 2018;**13**(2):92-109

[13] Hossain SA, Hossen M. Statistically processing of different sounds of vocalizing fish and mammals to estimate their population size with two acoustic sensors. Marine Technology Society Journal. 2019b;**53**(4):68-80

[14] Hossain SA, Hossen M. Population size estimation of chirp and grunt generating fish and mammals using cross-correlation based technique with three acoustic sensors. Journal of Ocean Engineering and Science. 2019c:183-191

[15] Hossain SA, Hossen M. Biomass estimation of a popular aquarium fish using an acoustic signal processing technique with three acoustic sensors. In: 2018 International Conference on Advancement in Electrical and Electronic Engineering (ICAEEE). IEEE; 2018. pp. 1-4

[16] Rana MS, Anower MS, Siraj SN, Haque MI. A signal processing approach of fish abundance estimation in the sea.

In: 9th International Forum on Strategic Technology (IFOST). IEEE; 2014. pp. 87-90

[17] Hossain SA, Hossen M. Selection of the optimum estimation parameter in cross-correlation based fisheries stock assessment technique. Journal of Ocean Technology. 2020;**15**(2):105-119

[18] Hossain SA, Hossen M. Impact of underwater bandwidth and SNR on cross-correlation based population estimation technique of fish and mammals. Underwater Technology. 2019d;**36**(2):19-27

[19] Hossain SA, Hossen M. Error calculation in cross-correlation based population estimation technique of fish and mammals. Acoustical Science and Technology. 2019e;**40**(6):402-405

[20] Hossain SA, Hossen M. Impact of dispersion coefficient on cross-correlation based population estimation technique of fish and mammals. Journal of Ocean Technology. 2019f;**14**(3):79-92

[21] Hossain SM. Cross-correlation based acoustic signal processing technique and its implementation on marine ecology [Doctoral dissertation]. Khulna, Bangladesh: Khulna University of Engineering & Technology (KUET); 2018

[22] Anower MS. Estimation using cross-correlation in a communications network [PhD dissertation]. Australian Defence Force Academy; 2011

[23] Hossain SA, Mallik A, Arefin MA. A signal processing approach to estimate underwater network cardinalities with lower complexity. Journal of Electrical and Computer Engineering Innovations (JECEI). 2017;**5**(2):131-138

[24] Hossain SA, Mallik A, Hossen M. An analytical analysis on fish sounds. Akustika. 2019;**33**:15-23

[25] Parmentier E, Vandewalle P, Frederich B, Fine ML. Sound production in two species of damselfishes (*Pomacentridae*): *Plectroglyphidodon lacrymatus* and *Dascyllus aruanus*. Journal of Fish Biology. 2006;**69**(2):491-503

Chapter 3

Diving as a Scientist: Training, Recognition, Occupation - The "Science Diver" Project

Alexandros Tourtas, Kimon Papadimitriou,
Elpida Karadimou and Ralph O. Schill

Abstract

Conducting scientific work underwater is a challenging endeavor. From collecting samples to protecting underwater cultural heritage sites scientific divers need to address issues concerning scientific methodology, diving safety, professional acknowledgement, training, legal implications etc. All of these matters are handled in different ways depending on factors like region, organizations involved, legal framework, diving philosophy etc. producing a diverse framework on scientific diving as a distinct type of underwater work. The ScienceDIVER project's main objective is to study and analyze this fragmented landscape, in order to provide insight and suggestions towards a commonly accepted framework that will promote scientific diving as a means of forwarding knowledge both within the scientific community and its interaction with the public.

Keywords: scientific diving, dive training, professional diving, diving legislation, diving safety

1. Introduction

For more than a century now the underwater world has yielded priceless information on a variety of scientific disciplines. Whether it is the amazing mechanism and the impressive cluster of bronze sculpture from the Antikythera Shipwreck [1] or the valuable measurements on biodiversity and how climate change affects the ecosystem (among others [2]), the data that derive from underwater projects enrich our perception of the world daily and significantly. Going through this long list of underwater endeavors it becomes evident that dive-based research is considered a valuable tool in scientific progress. New data along with new methodologies spring out of the challenging underwater environment enhancing scientific processes and results. Moreover, diving for scientific purposes is also considered nowadays a substantial part of professional development for scientists that want to expand their horizon or excel through the development of specialized skills and expertise. Thus, it has become part of a growing business sector that combines the scientific world with the maritime industry. Established terms such as *Blue Growth* and *Blue Economy*[1] or even recently emerged ones such as *Blue Science*[2] reflect the dynamic

---

[1] https://ec.europa.eu/maritimeaffairs/policy/blue_growth_en

[2] https://www.euromarinenetwork.eu/activities/blue-science-blue-growth

environment that the combination of several scientific fields can create working with the relevant public or private institutions, in order to promote social and financial development.

However, a career path to scientific diving (SD) is not clearly evident to those who seek to follow it, either students or scientists who want to forward their research underwater. The reason is probably the existence of multifarious ways in which different parts of the world or different established frameworks approach scientific diving as a part of their activity. The relevant landscape is chaotic not in the sense that it is totally absurdum of course, rather than in the mathematical sense of the term meaning that it has many variables that sometimes interact and other times remain idle, creating an unstable model for the harmonization of procedures and accreditation. Differences in philosophy that span from minor dissimilarities in definitions [3–9], to completely unlike and sometimes controversial approaches on features like health and safety [3, 5, 7, 10–13], remuneration and professional acknowledgment. That being said, there are of course established frameworks that do work on a regional, national or even continental level that have been developing for decades (among others [14–16]). Yet, since science leads the way in joining multi-backgrounded people for the promotion of knowledge and has in a way already achieved a global understanding on methodology and procedures, one should expect or even better strive towards the creation of a common framework for the scientific diving community as well, so as to promote research and expand international collaboration. United Nations' declaration of the decade 2021–2030 as the *Decade of Ocean Science for Sustainable Development*[3] is for example a great opportunity for nations to work together in order to generate the global ocean science needed to support the sustainable development of our shared oceans. Scientific diving could be a major device in providing an effective framework for the promotion of *Ocean Literacy*[4] and the enhancement of interaction between science and the public.

All the above generated the idea of a focused research on this particular field that would provide insight on effective ways for the creation of a unified scientific diving framework. The project "ScienceDIVER: Cross-sectoral skills for the blue economy market"[5] started in November of 2019 and comprises the joint effort of three Universities (Aristotle University of Thessaloniki, Greece - University of Calabria, Italy - University of Stuttgart, Germany), a research Institution (DAN Europe) and three companies representing the advisory maritime industry (Atlantis Consulting, Greece – envirocom, Germany – Marine Cluster Bulgaria). It is funded by the European Maritime and Fisheries Fund (Blue Economy 2018[6]) and its main objective is to support the development of blue and smart cross-sectoral skills, in order to meet the evolving needs in the labor market of Blue Economy. By building solid -long lasting- collaborations and structures between academia and industry it aims to offer standardized training and clear career pathways to diving scientists within the European Union. The project is structured in three phases. Firstly, there is the mapping of the relevant landscape and the assessment of needs. Subsequently the consortium will develop tools for the promotion of the project's objectives and lastly there is the testing phase and the provision of viable solutions along with the final results.

---

[3] https://en.unesco.org/ocean-decade; https://oceandecade.org/

[4] https://oceanliteracy.unesco.org/

[5] https://www.sciencediver.eu/

[6] https://ec.europa.eu/info/funding-tenders/opportunities/portal/screen/opportunities/topic-details/emff-03-2018

2. Methodology

This chapter was produced based on data from the first phase of the project. For all the reasons stated above mapping the landscape is a challenging endeavor. In order to approach the subject, the work was divided in five separate tasks. First one was mapping the stakeholders. Since it is nowadays widely accepted, both for professional and social reasons, that knowing the people involved is essential for the success of any management plan, identifying the stakeholders became a priority. The second task was in essence an expansion of the first one, since it comprises focused interaction with selected stakeholders such as competent organizations and policy makers. Both of the above two tasks were used as tools for the production of the following. Third task was to map the training framework concerning scientific diving. Task number four aimed at providing a view on the relevant legal framework, whereas the last one was focused on presenting matters of professional acknowledgement.

Data were gathered from various sources. Bibliography seems to be limited on the specific topic [17–20], since the bulk of scientific diving literature is mostly devoted to the presentation of projects or dedicated to specialized procedures, e.g. diving physiology or hazmat diving, rebreather diving etc. (for a list of indicative publications see [21] but also [22–25]) and not on theoretical matters concerning overall methodology and processes. This chapter is actually a way of contributing to this area of interest by disseminating the results of the project's survey. Most of the material used was taken from official texts provided by organizations that are either focused on or adjacent to scientific diving. Corpora with scientific diving guidelines and standards, various manuals (training, guidelines etc.), educational material etc. [3, 5, 7, 10, 12, 13, 26] that are provided by these organizations entail basic information concerning procedures, prerequisites, certification etc. but also reveal, although not straightforward the philosophy behind the choices on these matters. In other words, analyzing the data one may find clues on the various factors that create the general context that has led to those choices. On a more direct approach, once the stakeholders were identified and assessed, a series of interviews took place with various key players, in order to receive some insight to specific issues. In most cases, in order to keep a coherent approach, predefined questionnaires were used in the communication. Additionally, questionnaires were used also on specific subjects (e.g. citizen science and SCUBA diving) aimed at gathering data from a larger base like for example the recreational diving community. Beyond that, a lot of information was produced by online sources, such as official (and unofficial) websites (among others [2, 14–16, 27, 28]), social media etc. which were critically assessed and provided a more popular aspect to the research than the sterilized image official documents or official representatives do.

Moving from the greater context to more specific ones and trying to keep the overall picture while focusing on more specific areas the study was carried out on several levels. Starting from a global perspective, the first level's aim was to provide an overview of the situation at various parts of the world organizing big clusters. Most of these clusters were representative of continental regions (Europe, America, Australia, Asia, Africa) and the analysis was carried out on a superficial level providing, as stated above, an overview of the situation. More extensive was the study in selected areas. Europe was obviously the main focal point, however more detailed analysis was produced as well for other regions that were considered to have an important background in scientific diving such as North America (USA and Canada), Australia and New Zealand, South Africa, certain Asian countries etc. Deeper examination was then decided to be put forward in five "focus countries" in

the EU (Croatia, France, Germany, Greece, Italy) for them to be used as "case studies" keeping in mind that these would also probably serve as testing countries in the
implementation phase of the project.

Since this is an ongoing project and data gathering never stops on one hand and
the data pool is considerably large on the other, there were certain assumptions
made, in order to go on with the process. Such assumptions would be that the data
provided by official websites are indeed valid and updated, that the data provided
by various contact persons are true and correct and that in all cases there may be
more information available that has not been located yet during bibliography or
online searches, or has not been mentioned by the interviewees.

3. Training framework

The study of the training framework was deemed to be extremely important,
since it is the backbone of every diving scheme. It entails all the theoretical and
methodological information of each diving framework and has been designed to
transmit this information from one person to another, thus it is designed to be com-
prehensible and coherent. The scope of this task was to provide an overview of the
scientific diving training landscape and through comparative and analytical tools to
offer some insight on the various approaches that are taken spotting either common ground or indisputable differences. In short to provide a critical map of this
entangled network. The objectives set in order to provide this result were firstly to
make a list of all the official diving courses related directly or indirectly to scientific
diving, to study them and produce some analytical/comparative interpretations and
discuss them, in order to come up with relevant conclusions.

A total of 33 diving courses were presented and studied [28–50]. The data gathered comprised basic information about the training agency and the specific course
(title, weblink, short description), more detailed data concerning the content and
the learning objectives (theoretical knowledge and practical skills), some data
about logistics (examination, region, prerequisites, training material, certification
requirements and contact person information, i.e. name, position/assignment,
contact info). In addition to all the above, a quantitative attribute was placed
expressing the relevance of the specific course to scientific diving. The scale span
from 3 (max) which indicates a direct reference to SD in course title and description
to 1 (min) stating that there is no reference to scientific diving, although some of
the courses' content is adjunct to it, insinuating a low relevance. Relevance factor
index 2 represents the area between the two extremities with courses that although
not named as such, include references to scientific diving in their syllabus. The data
derived from all available sources (see above in methodology, p. 3).

Once the list was ready, the analytical phase of the study begun. For methodological reasons the following scheme "Training standards ->Training course ->
Certification -> Qualification/recognition" (**Figure 1**) was adopted in order to
be able to organize/categorize the data (training courses, organizations, learning objectives, material, prerequisites etc.). More specifically, four [4] criteria
categories were recognized, i.e. (a) Prerequisites (input), (b) Technicalities

Figure 1.
Training scheme as approached by the study.

(durations, costs, number of participants, etc.), (c) Learning objectives (knowledge and skills) and (d) Certifications (outputs). The above criteria were selected in order to allow the description of the training courses on the context of "pathways", originating from an entry point (Prerequisites), passing through a fulfillment phase (Learning Objectives) and resulting to a destination point (Certification) (**Figure 2**). Technicalities were considered as a complementary set of information for the logistics of each training course. It was later decided to let this part aside because the complexity of adjusting factors (region, currencies and local financial context, flexible training timelines etc.) made it impossible to gather usable data for comparative analysis and moved beyond the scope of the intended study.

From analyzing the gathered information that comprises a variety of training courses, the study has concluded to six [6] recognizable qualification systems, which are directly related to scientific diving: (a) American Academy of Underwater Science (AAUS) /Canadian Association of Underwater Science (CAUS), (b) Australian Diver Accreditation Scheme (ADAS), (c) Confédération Mondiale des Activités Subaquatiques (CMAS), (d) European Scientific Diving (ESDP), (e) Global Underwater Explorers (GUE), (f) Health and Safety Executive (HSE). Apart from CMAS and GUE the other four (AAUS, ADAS, ESDP and HSE) are not training agencies, but are providing qualification. These organizations provide standards for the creation of training courses by their members (e.g. for AAUS see [51]). Comparison was held on three levels (see **Table 1**) following the scheme on **Figure 2**.

The first one was about prerequisites. Features that were considered were the entry level – diving certification needed for a diver to begin scientific diving training, administrative matters such as current professional status, age limits, nationality issues etc., medical requirements, swimming proficiency and watermanship standards. It was also noted if and what kind of entry exams are required and whether there is a need to prove experience through the number and the type of previously logged dives. Summarizing the results, it seems that there is a common threshold concerning existing dive certification. All of the systems require a degree of recreational diving status (whether basic e.g. Open Water Diver or novice e.g. CMAS **, Rescue Diver or equivalent) in combination with Basic Life Support capabilities (e.g. CPR, first aid, defibrillation, oxygen provision). Most of them require medical examination and again all of them require a number of logged dives to prove some kind of experience, although the number and type of dives fluctuates from a minimum of 25 dives of any type to more specific demands like dive planning, participation to science projects etc.

The second level of analysis focused on learning objectives. The idea here was that if one breaks down the complex structure of these training systems, the basic elements that they are made of are the "learning objectives". Kind of like the genes in an organism. The selection/combination of learning objectives, either those referring to theoretical knowledge, or those that have to do with practical skills is essential since they are the building materials of the training courses and this process reflects their scope and objectives. Breaking down the courses and analyzing the basic themes on which they are based resulted in a comparative list containing

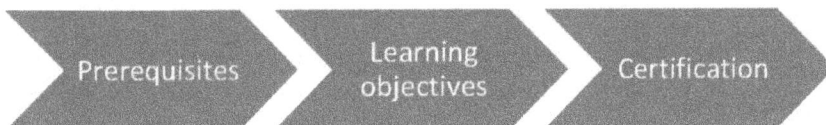

Figure 2.
Training "pathway".

	AAUS	ADAS	CMAS	ESDP*	GUE	HSE
PREREQUISITES						
ENTRY LEVEL (CERTIFICATION)	•	•	•	•	•	•
Autonomous diver". ISO 24801-2:2007 (e.g. Open Water Diver).	•	•				•
CMAS** or equivalent (e.g. Rescue Diver)			•	•	•	
EFR						•
ENTRY EXAMS	•	•				
ADMINISTRATIVE	•				•	
MEDICAL	•	•		•	•	•
SWIMMING / WATERMANSHIP	•	•			•	•
EXPERIENCE	•	•	•	•	•	•
Logged dives	•	•	•	•	•	•
Check dive	•					
LEARNING OBJECTIVES						
DIVE SAFETY	•	•	•	•	•	•
PROJECT MANAGEMENT	•	•	•	•	•	•
SCIENTIFIC METHOD	•		•	•	•	•
DATA RECORDING & HANDLING	•		•	•	•	•
Methods and Techniques	•		•		•	•
Mapping	•		•	•		
Data Management			•	•	•	
UW Imaging	•		•	•	•	
LEGAL ASPECTS	•	•	•	•	•	•
DIVE THEORY	•	•	•	•	•	•
DIVE MODES	•	•	•		•	•
SEAMANSHIP	•		•		•	•
SPECIAL CONDITIONS	•					•
SPECIALIZED EQUIPMENT	•		•	•	•	•
Full face mask	•					
Dry Suit	•					
Communications	•		•			•
Dive Propulsion Vehicle (DPV)	•					
SMBs/Lift Bags	•		•	•		
Line Reels	•					
Compressors			•			•
OTHER TOPICS	•		•	•	•	
CERTIFICATION						
TITLE	•	•	•	•	•	•

	AAUS	ADAS	CMAS	ESDP[*]	GUE	HSE
Scientific Diver	•		•	•	•	
Professional Diver						•
Occupational Diver		•				
RECIPROCITY	•	•				•
DISCIPLINES	•	•	•	•	•	•
Oceanography	•	•	•	•	•	•
Archaeology	•	•	•	•	•	•
Biology	•	•	•	•	•	•
Ecology	•	•	•	•	•	•
Geology	•	•	•	•	•	•
Engineering	•	•	•	•	•	•
Media						•

[*](KRISTINEBERG MARINE STATION).

Table 1.
Comparative analysis of scientific training schemes.

the following recognizable features: (a) dive safety, (b) project management, (c) scientific method, (d) data recording & handling (methodology, mapping, data management, u/w imaging), (e) legal aspects, (f) dive theory, (g) dive modes (e.g. SCUBA, CCR, SSD), (h) seamanship, (i) special conditions (e.g. chamber, night, deco dives), (j) specialized equipment (e.g. full face mask, dry suit, communications, DPV, lift bags, line reels, compressors), (k) other topics (e.g. u/w navigation, search methods, video systems). Through that process it became obvious that there are certain topics that are common to all training schemes such as dive safety, project management, legal aspects and dive theory. Moreover, it is evident that most of the systems (five out of six) provide also training on scientific methods, data recording and handling, dive modes and specialized equipment. Although there are differences on the extend and the ways each topic is approached, the aforementioned learning objectives seem to define the content of the term scientific diving as far as training is concerned.

The final level of analysis referred to the output of the whole process and more specifically to the provided certification and its acknowledgement. An important feature was the title granted after the completion of the training, due to the fact that it is absolutely related to its training systems' approach. Thus, along with the obvious Scientific Diver title, one comes across the terms: professional diver (HSE) and occupational diver (ADAS). This comes as no surprise of course, since the debate whether scientific diving belongs or not to the greater professional/occupational diving scheme is an old one and still raging. Moreover, the fragmentation of the scientific diving landscape also results in certifications not being globally recognized, although reciprocity arrangements are becoming more and more common between organizations, at least in terms of training. Lastly, it seems that the most common scientific areas where a scientific diving certification comes to fruition are oceanography, archaeology, biology – ecology, geology, engineering and as an adjacent field, media production for scientific purposes.

On a larger scale and beyond the three levels of analysis presented here, there are some other notable issues that were raised during the study. The fact for example that GUE and CMAS, the two systems that are also training organizations are preferable for individual training, since the rest require either a connection with a scientific institution or a professional status. These limitations also result in a tendency for gaps in the existing qualification systems to be covered by courses provided by training organizations (e.g. recreational diving agencies and universities). In other words, when none of the presented schemes is an option, which is quite common in many countries that have either a poor or no scientific diving framework at all, recreational diving courses with learning objectives related to scientific diving or short seminars organized by universities or relevant to the subject organizations seem to compensate for the lack of officially recognized certification. Lastly, a topic of an ongoing discussion that seems to be very interesting refers to the role of volunteers, amateurs and citizen scientists in research projects. There are courses for example, like the ones organized by the Nautical Archeological Society (NAS) in the UK [46] that have been educating citizens in underwater archaeology for decades now. Citizen science is rapidly developing in other scientific areas as well, most prominently in ecology and underwater biology, requesting from recreational divers to submit data to scientific projects [52–54]. It is evident that the results of this discussion affect the form of SD training schemes and the option of introducing the act of raising awareness, educating people and finally certifying them officially to participate in scientific diving projects could be a big part of scientific diving training in the future.

4. Legal framework

The importance of the legal framework surrounding scientific diving activities is in a way self-evident. It provides the solid base on which theory becomes practice and becomes -in the framework of this study- the liaison between training (i.e. structuring the method) with professional acknowledgement (i.e. applying it on the field). The diverse landscape that has been repeatedly mentioned so far is obviously reflected on legal matters as well. Although certain steps have taken place the last couple of decades, the present legal framework is still either pretty complex or insufficient (if nonexistent) making scientific diving a difficult task in terms of standardization, insurance, mobility (reciprocity) of scientific divers etc. Thus, it was decided that an overview of the legal framework along with an analysis of selected features or specific regional characteristics was absolutely necessary in order to comprehend the present situation, provide insights and produce suggestions for optimization.

The legal study was carried out following the basic methodology of the project, collecting data from all (mainly official in this case) available sources and communicating with stakeholders (e.g. diving industry, policy makers, scientific diving institutions) focusing on the legal aspect of their organization and activity. The data were organized and analyzed in such a way, so as to elucidate, if possible, the gray areas and identify the gaps among the various legal structures that have been built around scientific diving, in order of course to pave the way towards unification or at least cooperation.

The study comprises three basic parts. The first one focuses on definitions. Recognizing the importance of understanding the meaning of words and how much they represent the general principles of the organization issuing them on one hand and affect the implementation of each scheme on the other, it was deemed critical to have a more detailed look into them [3–9]. A list of definitions was created for comparative reasons containing the various interpretations that organizations involved in scientific diving apply to fundamental terminology. Besides the intuitively related terms, such as *scientific diving, scientific diver, science diver, scientific*

diving instructor, the list covers other major areas of diving activity such as commercial, occupational and recreational diving with the relevant terms, diving modes
(e.g. technical diving, surface supply, mixed gas diving, CCR) as well as other forms
of diving activity that have some kind of relevance with the investigated subject (e.g.
police and military diving, rescue procedures). Looking at the results one can see
that although the differences are minor the divergence further down the road is in
certain cases substantial creating a hiatus. However, the basic definitions are more
or less the same concluding that whatever the professional status or the dive mode
scientific diving is defined by (a) its scientific purpose and (b) its scientific method.

The second part of the legal study examines the international framework
covering scientific diving. Analyzing the current situation in various countries
the overall picture is again fragmentary. Differences occur on several levels, such
as for example the maturity of the legal framework itself. Certain countries (e.g.
Australia and New Zealand, Canada, South Africa, UK, and USA) have a long
tradition in occupational diving which means that they have an established legal
framework. On the contrary there are several countries, where scientific diving is
not mentioned at all in legal texts, even though there is a scientific community that
tries to regulate itself through recognized recreational diving schemes. In summary,
one could suggest that when not directly addressed the topic of scientific diving is
most often covered by labor laws or laws regulating maritime affairs. Another level
of differentiation includes means of authority and enforceability. In countries that
do have an established framework, authorities follow a scheme with four cooperat-
ing parts (**Figure 3**): (a) Statutes laws that are at the top of the legal hierarchy and
are created by a legal body. (b) Regulations. A regulation is the second step in the
hierarchy of law. Regulations have the force of law which is made by an executive
authority under powers delegated by primary legislation. (c) Codes of Practices.
Codes of Practices do not have the force of the law. They provide guidance from the
regulator on how to comply with requirements and obligations under work health
and safety laws and regulations. They can influence court proceedings under the
health and safety laws and regulations. (d) Standards. They do not have legal force,
but can similarly be used to establish norms for certain classes of diving. These are
voluntary consensus documents, which, although not automatically a legal document, may be incorporated into legislation by reference or used in private contracts
as a set of specifications and procedures. Thus, besides national legislations the
study included official texts issued by scientific diving organizations, such as the
AAUS, ESDP, EUF (European Underwater Federation) etc. [3–5, 10–13, 26, 55–58].
This leveled structure means that both the formation and the enforcement of the
relevant legislation is distributed to several stakeholders rather than being a solemn
responsibility of the state. Following a bottom up approach, organizations need to
look after the validity of their procedures according to standards and code of practices they helped to create, collective organisms and government departments have
to check and modify regulations whereas national administration is responsible for
the creation, modification and enforcement of the statutory laws. Of course, that's a
scheme that works in countries that do have a mature legal framework.

The third part of the legal study was based on more thorough research covering
the focus countries (i.e. Croatia, France, Germany, Greece and Italy). Specified

Figure 3.
The four levels of legal engagement.

research topics were investigated in order to provide further insight into the matter. Since it would be lengthy endeavor to present the content of the research here, it would suffice to say that the results were more or less anticipated and the picture drawn was not at all different from the one described in the previous paragraph. What is interesting enough is than in none of the focus countries the legal framework is specifically directed at scientific diving. Running the gamut there are (a) professional diving environments like the one in France regulating matters of scientific diving and addressing the matter on an occupational basis according to labor legislation, (b) lighter or less structured frameworks that are more related to scientific diving yet less official (e.g. Germany, Croatia, Italy) and (c) completely absent frameworks, like the case of Greece that has minimal scientific diving reference in official texts.

Concluding and keeping in mind that the scope of the project focuses on the European level, one could definitely suggest that there is a clear need for clarifications and legal framework improvements concerning scientific diving for the vast majority of countries and an effort to agree on basic principles in order to unify the regionally based arrangements that have been running so far. Especially within the European Union the tools for harmonization on such matters already exist and through this framework the scientific diving community can strive for a widely accepted legal structure.

5. Professional acknowledgement

Same as training and legal aspects, professional acknowledgement of scientific diving is a complex matter. An introductory statement could be that on a global scale scientific diving is rarely recognized as a distinct professional activity. However, since this statement is quite vague, the professional acknowledgment study aimed at providing a more accurate assessment of the situation. Presenting the international *status quo* and then focusing on particular regions and situations produced an evaluation of the degree of professional acknowledgment based on relevant data and expert opinions. In order to organize and analyze the data a list of selected criteria was created based on (a) definitions (e.g. scientific diving purpose, scientific methodology), (b) legal framework (e.g. official recognition by the state, relevant legal framework), (c) remuneration (occupational aspect) and (d) reciprocity (regional and institutional).

In general, one can suggest that the status of the relevant legal framework is directly affecting professional acknowledgement. Thus, in regions with an established legal framework professional recognition is much more developed that in the case of regions with a lack of or an insufficient legal framework. In the latter, recognition of scientific diving credentials becomes unofficial and is regulated on an organizational level e.g. by institutions or the private sector according to circumstantial needs.

An interesting outcome that derived from the study was the notion that scientific diving in practice comprises two interacting parts. The first one is obviously the scientific method used to acquire the information from the underwater environment. Disciplines like archaeology, biology, engineering, geology, medicine, oceanography, meteorology have for a long time now already developed methodological tools designed for underwater work. The second one refers to the diving aspect and is a cluster of techniques (or dive modes) that have to do with all the subsidiary activity that needs to take place during an underwater scientific project. Deep diving, heavy lifting, setting up the survey area, scientific equipment maintenance, recovering of artifacts and samples and other tasks that require a variety of skills which range from basic recreational diving skills to more demanding diving procedures, such as surface supply, mixed gases, lift bags handling, power tools etc. The first part is easily defined and accepted, as stated previously in this text. The second part though is actually an area of debate, since it involves tasks that are traditionally connected to commercial diving [3].

Another point of interest is the distinction between scientists that organize their own scientific diving projects as part of their wider research and those that are directly employed by institutions to work specifically for underwater scientific work. In other words, scientists that also dive and scientific divers explicitly hired for this purpose. This distinction refers to the occupational nature of the underwater work performed and becomes part of the relevant discussion. Once again there are cases where the distinction becomes difficult and definitions about occupational diving like the one used by ADAS [4] try to provide clear solutions stating that "*Diving in the course of employment (irrespective of whether or not diving is the principal function of employment or merely an adjunct to it) and comprising all diving work carried out as part of a business; as a service; for research; or for profit.*" is occupational. Of course, not everybody agrees with this statement and the debate lingers.

Another aspect of the occupational nature of scientific diving is remuneration. Information about scientific divers' remuneration is difficult to acquire due to the multifactorial nature of wages (depending on region, legal framework, professional status etc.) and the sensitivity of personal data involved. A general conclusion to be made though is that payment can be produced either in a direct way, in the form of a salary for scientific diving services or indirectly as compensation, allowance, supplementary payment etc. which aligns with the aforementioned distinction between scientists that dive in the scope of their work or scientific diver explicitly hired for that purpose.

Lastly, insurance - wise when things are not specifically regulated by the occupational diving framework and insurance is not provided by the state (public insurance), most of the countries recognize either DAN insurance [59] or other relevant occupational insurance schemes form the private sector. Accident insurances are offered by various insurance providers. However, they only cover the costs of treatment, which can of course be very cost-intensive. Much more important are the benefits that may be provided by the social insurance schemes, e.g. in the case of occupational disability due to an accident.

6. Conclusions

Summarizing results and discussion, here are some final thoughts on the subject. From a starting point that unification is a good thing and will promote the interests of the scientific diving community, the fragmentation that the study of today's landscape reveals needs to be addressed on certain fundamental issues.

A major decision that needs to be made towards unification is related to the basic questions on the professional nature of scientific diving. Is scientific diving an occupational activity? Should scientific diving follow commercial diving practices? Should it abide by strong rules that provide a stable framework or should it serve a less rigid yet versatile framework that provides options for more contexts?

It is impossible if not irresponsible to provide answers hastily. The scientific diving landscape is at the moment fragmentary due to regional or professional micro-management or simply because it is still undeveloped or immature in many countries, even in some that do have noticeable scientific diving activity. Thus, the development of a widely accepted scientific diving framework (training, legal, professional) is not just a way to enhance reciprocity among the existing ones, but more importantly a way to promote scientific diving in general.

There is an ongoing effort for the creation of a World Scientific Diving Training Council, while in our area of interest, Europe, the ESDP tries to bring people from various countries and backgrounds together during conferences and workshops [60–62]. The ScienceDIVER project itself is proof that work is being done towards that goal. Hopefully these efforts will come to fruition soon and scientific diving will be sufficiently supported in the years to come.

This chapter follows the project's primary goal which is to promote a new holistic approach to the theory behind the formation of a scientific diving framework. As stated above, bibliography focused on the theoretical matters of scientific diving is limited. In most cases there are thoughts and insights among the lines of texts dedicated either to the presentation of underwater scientific work or to presentations of the current situation in specific regions. In this chapter we present the data gathered during the analytical phase of the project providing a wide view on the relevant landscape concerning what we believe to be the main pillars of this structure, i.e. training, legal framework and professional acknowledgement. Most importantly through the analysis of the various tasks performed, we provide our suggestion for a methodology addressing this multileveled and complex cluster of issues that we believe corresponds to the fundamental features of scientific diving. The various approaches expressed hitherto concerning these three major features have been discussed in the relevant sections in a way, so as to pinpoint the key issues of confrontation, such as the content of training courses (e.g. learning objectives and prerequisites), the various legal implications, safety and insurance policies, salary issues etc. Having all the above gathered in one study will serve as a base for the development of new scientific bibliography focused on the issues raised in this text. In other words, we hope that this text will serve as a trigger point for the production of new scientific literature of the same holistic nature, since the problems we face will only be efficiently handled through collaboration and convergence. In order to do so though we need more people and projects working on the issue. Whether the authors of these future study agree or not with the thoughts expressed in this text is of minor importance, since the discussion will have commenced.

Acknowledgements

The ScienceDIVER project is funded the European Maritime and Fisheries Fund (Blue Economy 2018). The data used in this chapter derive from deliverables produced by the project's partners. Contributions were made by Konstantinos Tokmakidis and Georgia Kalantzi from AUTh, Franz Brummer from UST, Fabio Bruno and Matteo Collina from UNICAL, Angelos Manglis, Themistoklis Ioannidis, Paschalina Giatsiatiou and Dimitra Papadopoulou from ATL, Salih Murat Egi and Guy Thomas from DAN, Anton Krastev, Rumen Stoyanov, Ilze Athanasova and Siyana Angelova from MCB.

Conflict of interest

The authors declare no conflict of interest.

Author details

Alexandros Tourtas[1*], Kimon Papadimitriou[1], Elpida Karadimou[1]
and Ralph O. Schill[2]

1 Aristotle University of Thessaloniki, Thessaloniki, Greece

2 ENVIROCOM, (www.envirocom.de), Tübingen, Germany

*Address all correspondence to: alextourtas@topo.auth.gr

IntechOpen

References

[1] Kaltsas, N., Vlachogianni, E. & Bouyia, P. (eds.) 2012. The Antikythera Shipwreck. The ship, the treasures, the Mechanism. Athens: National Archaeological Museum, 303 pp. ISBN-13: 978-9603860310

[2] Molinos, J.G., Halpern, B.S., Schoeman, D.S., Brown, C.J., Kiessling, W., Moore, P.J., Pandolfi, J.M., Poloczanska, E.S., Richardson, A.J. and Burrows, M.T., 2016. Climate velocity and the future global redistribution of marine biodiversity. Nature Climate Change, 6(1), pp.83-88.

[3] AAUS Standards for scientific diving manual [Internet]. 2019. Available from: https://aaus.org/diving_standards [Accessed: 2020-07-30]

[4] ADAS. Scientific diving [internet]. 2020. Available from: https://adas.org.au/careers/scientific-diving/ [Accessed: 2020-10-10]

[5] CMAS Standard for scientific diver. CMAS Scientific Diver. CMAS Advanced Scientific Diver. CMAS Scientific Diving Instructor. CMAS Scientific Diving Confirmed Instructor [Internet]. 2020. Available from: https://www.cmas.org/science/standards [Accessed: 2020-10-15]

[6] Italian Association of Scientific Divers [Internet]. 2020. Available from: http://www.aioss.info/default_e.asp [Accessed: 2020-10-15]

[7] NOAA Diving Manual 5th Edition [Internet]. 2017. Available from: http://tecvault.t101.ro/NOAA%20Diving%20Manual.pdf [Accessed: 2020-07-30]

[8] Flemming NC, Max MD, editors. Scientific diving: a general code of practice. 2nd ed. UNESCO Pub.; 1997. 254 p. ISBN-13: 978-9231032509

[9] US DoL. Occupational Safety and Health Administration. Definitions. Available from: https://www.osha.gov/pls/oshaweb/owadisp.show_document?p_id=9978&p_table=STANDARDS [Accessed: 2020-10-15]

[10] HSE Diving at Work Regulations 1997 dated 13 February 2020 [Internet]. 2020. Available from: https://www.hse.gov.uk/diving/qualifications/approved-list.pdf [Accessed: 2020-07-30]

[11] NOAA Diving Standards and Safety Manual [Internet]. 2017. Available from: https://www.omao.noaa.gov/sites/default/files/documents/NDSSM%20Rev%2012%20Feb%202020.pdf [Accessed: 2020-07-30]

[12] Société de Physiologie et de Médecine Subaquatiques et Hyperbares de Langue Française. Société Française de Médecine du Travail. Prise en Charge en Santé au Travail Des Travailleurs Intervenant en Conditions Hyperbares. Deuxième edition. [Internet]. 2018. Available from: https://www.medsubhyp.fr/images/consensus_bonnes_pratiques_reglementation/Sant-au-travail-des-travailleurs-hyperbares-2018-v2.pdf [Accessed: 2020-07-30]

[13] Buone prassi per lo svolgimento in sicurezza delle attività subacquee di ISPRA e delle Agenzie Ambientali. [Internet]. 2013. Available from: https://www.isprambiente.gov.it/files/pubblicazioni/manuali-lineeguida/MLG_94_13_Buone_prassi_attivit_subacquee.pdf [Accessed: 2020-07-30]

[14] AAUS. Overview. [Internet]. 2020. Available from: https://www.aaus.org/ [Accessed: 2020-10-15]

[15] ADAS. Overview. [Internet]. 2020. Available from: https://adas.org.au/adas-overview/ [Accessed: 2020-10-15]

[16] HSE. Overview. [Internet]. 2020. Available from: https://www.hse.gov.uk/diving/ [Accessed: 2020-10-15]

[17] Ponti, M. 2012. Scientific diving: towards European harmonization, *International Journal of the Society for Underwater Technology* 30(4), pp. 181-182

[18] Sayer, M.D.J., Fischer, P. & Feral J.P. 2008. Scientific Diving in Europe: Integration, Representation and Promotion In: *Brueggeman P, Pollock NW, eds. Diving for Science 2008. Proceedings of the American Academy of Underwater Sciences 27th Symposium. Dauphin Island*, AL: AAUS, pp. 139-146

[19] Sayer, M.D.J. 2007. Scientific diving: a bibliographic analysis of underwater research supported by SCUBA diving, 1995-2006 *International Journal of the Society for Underwater Technology* 27(3), pp. 75-94

[20] Sayer, M.D.J. & Barrington, J. 2005. Trends in scientific diving: an analysis of scientific diving operation records, 1970-2004 *International Journal of the Society for Underwater Technology* 26(2), pp. 51-55

[21] Diving Publications. Scientific Diving. US EPA. Available from: https://www.epa.gov/diving/diving-publications [Accessed: 2020-10-15]

[22] Pollock, N.W., Sellers, S.H. & Godfrey, J.M. (eds.) 2016. *Rebreathers and Scientific Diving. Proceedings of NPS/NOAA/DAN/AAUS June 16-19, 2015 Workshop*. Wrigley Marine Science Center, Catalina Island, CA; 272 pp. ISBN: 978-0-9800423-9-9

[23] Kur, J. & Mioduchowska, M. 2018. Scientific Diving in natural sciences, *Journal of Polish Hyperbaric Medicine and Technology Society* 4(65), pp. 55-62 DOI: 10.2478/phr-2018-0024

[24] Balazy, P., Kuklinski, P. & Wlodarska-Kowalczuk, M. 2014. Scientific Diving in polar regions – the example of ecological studies at the Institute of Oceanology, Polish Academy of Sciences, *Journal of Polish Hyperbaric Medicine and Technology Society* 1(46), pp. 65-84

[25] Merkel, B.J. & Schipek, M (eds.) 2009. Research in Shallow Marine and Fresh Water Systems 1st International Workshop - Proceedings Freiberg Online Geology 22

[26] ESDP Common Practices for Recognition of European Competency Levels for Scientific Diving at Work European. Scientific Diver (ESD). Advanced European Scientific Diver (AESD). Consultation Document 1 (rev. 1) [Internet]. 2017. Available from: http://ssd.imbe.fr/IMG/pdf/esdp_consult_1_rev1_april_2017_sd_competency_levels_.pdf [Accessed: 2020-09-25]

[27] CMAS. Overview. [Internet]. 2020. Available from: https://www.cmas.org/ [Accessed: 2020-10-17]

[28] GUE Scientific Diver [Internet]. 2020. Available from: https://www.gue.com/diver-training/explore-gue-courses/foundational/scientific [Accessed: 2020-07-30]

[29] ADAS Certification Conditions [Internet]. 2020. Available from: https://adas.org.au/adas-certification-conditions/ [Accessed: 2020-07-30]

[30] ADAS Certification Overview [Internet]. 2020. Available from: https://adas.org.au/need-occupational-diving-certificate/ [Accessed: 2020-07-30]

[31] ADAS Training [Internet]. 2020. Available from: https://adas.org.au/training / [Accessed: 2020-07-30]

[32] ADAS Training Courses [Internet]. 2020. Available from: https://adas.org.au/training-courses/ [Accessed: 2020-07-30]

[33] ADAS Training Schools [Internet].
2020. Available from: https://adas.
org.au/training-schools [Accessed:
2020-07-30]

[34] CMAS Scientific & Sustainability
Committee. Non-professional CMAS.
Scientific Specialty Courses (SSC).
ADMINISTRATIVE TEXT [Internet].
2018. Available from: https://www.
cmas.org/science/standards [Accessed:
2020-10-15]

[35] CMAS Scientific & Sustainability
Committee. Underwater Geology
Course (UGC) [Internet]. 2018.
Available from: https://www.cmas.
org/science/standards [Accessed:
2020-10-15]

[36] CMAS Scientific & Sustainability
Committee. Underwater Cultural
Heritage Discovery Course (UCHDC)
[Internet]. 2018. Available from: https://
www.cmas.org/science/standards
[Accessed: 2020-10-15]

[37] CMAS Scientific & Sustainability
Committee. Underwater Archaeology
Course (UAC) [Internet]. 2018.
Available from: https://www.cmas.
org/science/standards [Accessed:
2020-10-15]

[38] CMAS Scientific & Sustainability
Committee. Ocean Discovery Course
(ODC) [Internet]. 2018. Available
from: https://www.cmas.org/science/
standards [Accessed: 2020-10-15]

[39] CMAS Scientific & Sustainability
Committee. Marine Biology Course
(MBC) [Internet]. 2018. Available
from: https://www.cmas.org/science/
standards [Accessed: 2020-10-15]

[40] CMAS Scientific & Sustainability
Committee. Freshwater Biology Course
(FBC) [Internet]. 2018. Available
from: https://www.cmas.org/science/
standards [Accessed: 2020-10-15]

[41] CMAS Scientific & Sustainability
Committee. Conservation Biology

Course (CBC) [Internet]. 2018.
Available from: https://www.cmas.
org/science/standards [Accessed:
2020-10-15]

[42] CMAS Scientific & Sustainability
Committee. Advanced Underwater
Geology Course (AUGC) [Internet].
2018. Available from: https://www.
cmas.org/science/standards [Accessed:
2020-10-15]

[43] CMAS Scientific & Sustainability
Committee. Advanced Underwater
Archaeology Course (AUAC) [Internet].
2018. Available from: https://www.
cmas.org/science/standards [Accessed:
2020-10-15]

[44] CMAS Scientific & Sustainability
Committee. Advanced Marine Biology
Course (AMBC) [Internet]. 2018.
Available from: https://www.cmas.
org/science/standards [Accessed:
2020-10-15]

[45] CMAS Scientific & Sustainability
Committee. Advanced Freshwater
Biology Course (AFBC) [Internet].
2018. Available from: https://www.
cmas.org/science/standards [Accessed:
2020-10-15]

[46] NAS Education. International
System [Internet]. 2020.
Available from: https://www.
nauticalarchaeologysociety.org/
Pages/Category/international-system
[Accessed: 2020-07-30]

[47] NOAA Science Diver Qualification
Requirements [Internet]. 2015. Available
from: https://www.omao.noaa.gov/
sites/default/files/documents/0306%20
-%20NOAA%20Science%20Diver%20
Qualification%20Requirements.pdf
[Accessed: 2020-07-30]

[48] NOAA Scientific Diver Training and
Certification Requirements [Internet].
2015. Available from: https://www.
omao.noaa.gov/sites/default/files/
documents/0313%20-%20NOAA%20

Scientific%20Diver%20Training%20
and%20Certification%20Requirements.
pdf [Accessed: 2020-07-30]

[49] University of Gothenburg. 5th
International German - Swedish - Finish
European Scientific Diving (ESD)
Training Course 2019 [Internet].
2019. Available from: https://www.
euromarinenetwork.eu/system/
files/2019/5th%20ESD%20Training%20
Course%202019.pdf [Accessed:
2020-07-30]

[50] Coordination of Scientific Divers
of Croatia. Scientific Diving Course.
Spring 2015. International Centre
for Underwater Archaeology Zadar
SUB [Internet]. 2015. Available from:
http://icua.hr/images/stories/natjecaji/
Scientific%20Diving%20Course%20
ICUA%202015.pdf [Accessed:
2020-07-30]

[51] AAUS Members [Internet].
2020. Available from: https://www.
aaus.org/AAUS/Members/Current_
Organizational_Members/AAUS/
Organizational_Members_Public.
aspx?hkey=8cfd3868-610c-4285-afc3-
57f0d5fa2c09 [Accessed: 2020-07-30]

[52] Cohn, J. P. 2008. Citizen Science:
Can Volunteers Do Real Research?
BioScience 58(3) pp. 192-197

[53] Dickinson, J., Shirk, J., Bonter,
D., Bonney, R., Crain, R.L, Martin,
J., Phillips, T & Purcell, K. 2012. The
current state of citizen science as a
tool for ecological research and public
engagement. *Frontiers in Ecology and the
Environment* 10(6), pp. 291-297

[54] Barnes, J.A. 2017. Public archaeology
is citizen science in Arkansas, *Journal
of Community Archaeology & Heritage*,
DOI: 10.1080/20518196.2017.1308296

[55] HSE Diving at Work Regulations
1997. HSE criteria for approval of
non-UK diving qualifications dated
9th July 2018 [Internet]. 2018.

Available from: https://www.hse.gov.
uk/diving/qualifications/non-uk-
diving-qualifications.pdf [Accessed:
2020-07-30]

[56] NOAA Scientific Training
and Proficiency Dives [Internet].
2015. Available from: https://www.
omao.noaa.gov/sites/default/
files/documents/0302%20-%20
Scientific%20Training%20and%20
Proficiency%20Dives.pdf [Accessed:
2020-07-30]

[57] NOAA Science Divers Performance
of Working Tasks [Internet]. 2015.
Available from: https://www.
omao.noaa.gov/sites/default/files/
documents/0303%20-%20Science%20
Divers%20Performance%20of%20
Working%20Tasks.pdf [Accessed:
2020-07-30]

[58] Model Code of Practice for Safe
Diving Operations. Prepared by the
UNESCO UNITWIN Network for
Underwater Archaeology [Internet].
2013. Available from: http://www.
underwaterarchaeology.net/
News/UNITWIN_Model_Dive_
Code_20190620_EN.PDF [Accessed:
2020-10-10]

[59] Divers Alert Network Europe.
Insurance. [Internet]. 2020. Available
from: https://daneurope.org/web/guest/
insurance [Accessed: 2020-10-16]

[60] ESDP Report 18th Plenary Meeting
European Scientific Diving Panel. 11
April 2017 | Station Marine d'Endoume
| Marseille [Internet]. 2020. Available
from: http://ssd.imbe.fr/IMG/pdf/18th_
esdp_meeting_report_vf.pdf [Accessed:
2020-07-30]

[61] ESDP Report. European Scientific
Diving Panel (ESDP) 19th Meeting – 26
October 2017 MUMM Brussels & Skype
[Internet]. 2017. Available from: http://
ssd.imbe.fr/IMG/pdf/esdp_-_minutes_
meeting_19_-_261017.pdf [Accessed:
2020-07-30]

[62] ESDP 20th Meeting – 04th May 2018 -University of Stockholm & Skype [Internet]. 2018. Available from: http://ssd.imbe.fr/IMG/pdf/esdp_-_minutes_meeting_20_-20180504.pdf [Accessed: 2020-07-30]

Chapter 4

Progressive Underwater Exploration with a Corridor-Based Navigation System

Mario Alberto Jordan

Abstract

The present work focuses on the exploration of underwater environments by means of autonomous submarines like AUVs using vision-based navigation. An approach called Corridor SLAM (C-SLAM) was developed for this purpose. It implements a global exploration strategy that consists of first creating a trunk corridor on the seabed and then branching as far as possible in different directions to increase the explored region. The system guarantees the safe return of the vehicle to the starting point by taking into account a metric of the corridor lengths that are related to their energy autonomy. Experimental trials in a basin with underwater scenarios demonstrated the feasibility of the approach.

Keywords: monocular active SLAM, vision-based navigation, dense mapping, feature-based SLAM, corridor network, path planning, guidance system, self-similarity, sunlight caustics

1. Introduction

Vision-based SLAM systems are much appreciated nowadays in a broad spectrum of applications in robotics, autonomous navigation, and guidance systems, in airborne, underwater, and terrestrial environments [1–4]. A global comparison shows that SLAM applications underwater are less abundant (see some state-of-the-art works) [5–10].

In particular, underwater navigation generally excludes any form of global positioning system such as in other environments, so vision is a sound alternative, provided visibility is good enough. On the other hand, in turbid waters it is possible, to some extent, to navigate at low altitude on the seabed. However, in these extreme cases, the texture of the seabed may appear voluminous, which implies visual occlusions and collision probabilities. Conversely, shallow waters, rapid changes in the natural illumination, for example, due to the sunlight caustic on the floor, shadows, and flashes, can seriously damage the photometric properties of the images. In such extreme cases, the estimation of the vehicle position can be so impaired with the possibility of vehicle loss.

These characteristics of the underwater environment pose major challenges to the success of navigation, especially if it takes place in unknown regions. These challenges relate not only to robust estimations of the vehicle position and surrounding cartography [11] but also to the ability of the guidance system to avoid possible collisions at low altitudes.

In this work we will present an approach for the autonomous exploration of unknown regions of the seabed by means of a navigating system connected to a decision-making process hosted in the submarine. This vision-based approach is called Corridor SLAM (C-SLAM), because it combines SLAM techniques with a strategy to build a corridor over the seabed. The original concept was described in [12]. This document presents a generalization of previous work centered on a robust network of corridors. A variant of the basic concept uses active SLAM for exploring and building a grid map (see recent works [13–15]), but they do not enclose in any way the concept of path optimality as here, which proposes continuous paths with viewpoints associated.

A relatively close idea to the concept of corridor can be found in the "teach and repeat" method, [16], to follow a path in unfamiliar outdoor environment, such as in the exploration on the surface of Mars by a rover. However, the basic method does not include an autonomous exploration, as the first path construction is given remotely by a human operator with the aid of a camera at the rover front. Once an a priori feasible path has been defined, the vehicle can "repeat" the path that was previously "learned."

The main objective of our approach is to make the right decisions to build robust pathways to explore and configure them in a network. At the same time, the system is able to bring the vehicle back to the starting point from any position in the explored area. The robustness properties of the network are conceived in the sense that there are preferential directions to explore and are such that one allows a successful return.

2. C-SLAM-based autonomous navigation

Figure 1 illustrates the complete structure of the proposal for an autonomous vision-based system. It consists of three nested feedback loops, called adaptive control loop, reference adjustment loop, and dense mapping loop. They are described in details below.

The adaptive control loop generates the control actions u necessary to achieve the prescribed reference path $\hat{\eta}_{ref}$ and the reference speed $\hat{\dot{\eta}}_{ref}$. Both are provided by the guidance system. The path following will attenuate the influence of sea currents

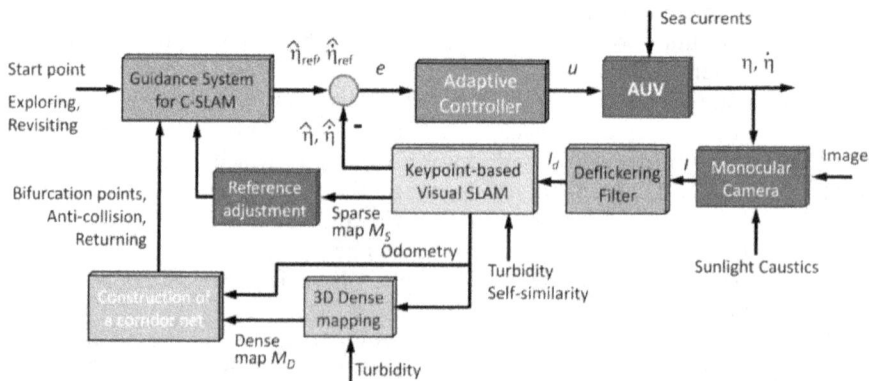

Figure 1.
C-SLAM structure. Three-loop-based approach to underwater exploration.

that push the vehicle off the corridor. The adaptive law circumvents the requirement to make available an AUV model which is generally complex due to its 6
degrees of freedom (see [17, 18]). Additionally, the action of marine currents and
a variable speed of the AUV make hydrodynamic resistance forces very difficult to
estimate and include in the dynamic model. For these reasons, controllers with
self-adjustment of coefficients and dynamics adaptation seem to be much more
effective than fixed controllers.

In the adaptive control loop feedback, there is the vision system that basically
takes images of the seabed with the vehicle in motion to estimate the position η and
rate $\dot{\eta}$ of the AUV. The estimates are termed $\hat{\eta}$ and $\hat{\dot{\eta}}$, respectively. The monocular
camera onboard takes images in a sequence called I. The line of vision is oriented
forward with a certain inclination towards the ground. In case of sunlight caustics
on the seafloor [19], the quality of the images may be significantly improved by an
online deflickering filter [20–22]. The filter attempts to retrieve the original photometric properties of the frame. Its output is the frame sequence I_d. A feature-based
SLAM algorithm estimates the position of the vehicle and a sparse depth map M_S of
the floor. Here, texture characteristics of the ground like self-similarity and water
turbidity will generally difficult the self-localization and map estimation. Therefore
they are considered as disturbances [23, 24].

The reference adjustment loop has the function of reformulating the references
$\hat{\eta}_{ref}$ and $\hat{\dot{\eta}}_{ref}$ according to a new scale factor of the estimated relief. This is necessary
due to the variability of the estimated sparse map when revisiting the corridor site.

The outer loop of the C-SLAM, namely, the dense mapping loop, is dedicated
to the construction of the corridor network, providing alternatives for bifurcating
the path during the exploration. The feedback consists of the navigation system,
a block for estimating a dense mapping M_D of the terrain, and a block for path
planning and bifurcation managing. These blocks will process the odometry information provided by the inner feedback in order to update the topography of the
corridors. The dense map helps the guidance system to implement the future path
to follow and to avoid collisions.

A typical navigation route consists in part of stretches between relevant points
or landmarks. For example, a stretch that links a starting point with a bifurcation
point, or one that covers an entire branch, or one between a bifurcation point and a
point of no return. During the exploration, the path is created step by step, in which
the vehicle direction is regularly changed towards preferential points that should
ensure the continuity of the future path.

According to the main objectives of the exploration, C-SLAM should allow a
successful return of the vehicle to the starting point from any distant point of the
network. The farthest points are called no-return points. In this case, it is assumed
that when the vehicle reaches a point of no return, it possesses at least half of its
energy autonomy [25–27]. It is therefore that C-SLAM must permanently estimate
points of no return and also ensure that there exists at least one way for the
comeback before losing the energy autonomy.

3. Corridor net

The network of submarine corridors is the composition of interconnected strips
on the seabed, which covers a certain explored region and allows navigation
between different points of it. Topographically the strips are sequences of recognizable sites on M_S and topologically by a branched network that is made up of
nodes and bifurcation points.

A node is characterized by a cluster of seabed characteristics. It has a position and a viewpoint associated with the local characteristics. These characteristics are represented by keypoints of the local scene. The position of the cluster is determined by its centroid.

In the exploration, C-SLAM has to decide at each moment the next step of movement that starts from the current node. Each step consists of the identification of the next node, the direction of its centroid, and the speed of navigation to bridge the gap between nodes. The detection and identification of a node is a process that can involve many steps, during which a node is tracked and classified among other nodes in a merit order list.

In order to choose a suitable node among many possible ones, the vision system evaluates the most outstanding characteristics of the scene and then decides the optimal direction towards the corresponding node. The optimization of this direction results from the maximization of a cost function of the quantifiable requirements. Similarly, the rate is properly determined based on a previously specified vehicle cruising speed. Thus, the speed can be regulated to go up and down according to the curvature of the path. This construction procedure is repeated in the next step permanently during the exploration.

Cluster tracking involves localization on both a scattered local map and a dense map. The estimation of distances between nodes of any network branch is decisive for the planning of revisits according to the available margins of autonomy of the vehicle.

It is worth noting that estimated distances on the maps are simply evaluated on the particular scale of the monocular SLAM method, which does not necessarily coincide with the actual scale. This would not represent an obstacle to C-SLAM as long as the estimated distance between the starting point and any point of no return on the network is exactly related to half the energy autonomy of the vehicle. Some approaches define another type of scaled topology, achieved by means of stereo vision or scanner [8, 10, 27].

Bifurcation points are commonly distributed along each branch. There are two different sets of bifurcation points. The first set includes divergent points at which the paths physically bifurcate. The second set covers the so-called open bifurcation points, which were identified and ranked in a list of suitable landmarks for a future deviation of the path. The ranking position is issued in accordance to a criterion of robustness.

While each new corridor is simply added to the network as a new branch, existing corridors are updated in each new revisiting. Each round trip implies that the vehicle starts with plenty of energy autonomy. This indicates the dimensions of the explored region. Clearly, with the deployment of multi-robots, the scan will be faster [28, 29] but will not necessarily cover a larger region than with this proposal.

Since the camera is monocular, safe return is practically implemented by guiding the vehicle in reverse. This is because the system might not identify all the nodes in the 3D landscape in the way back, i.e., when navigating from the opposite direction. However, with the camera in reverse, the system is able to recognize the seabed features on the path straightforwardly. In reverse motion, all nodes are revisited but from back to front.

4. Construction of a corridor net

Before starting a round trip in the network of corridors, C-SLAM must make the decision to advance the exploration or revisit explored branches of the network. Commonly, both decisions are involved with the intent to expand the explored region. In addition, the system defines a suitable ground altitude for navigation with the end of achieving good visibility conditions.

For the next discussion, **Figure 2** is taken as support.

The first corridor and each new section of the corridor network are constructed according to an optimum criterion. This criterion takes into account the most outstanding characteristics of the seabed in the form of keypoints (shortened KP), which can be categorically recognized over time and preferably from different points of view.

Autonomous navigation begins with a commissioning phase (CP) to adjust the controller and initialize SLAM techniques for scattered and dense mapping. To achieve these objectives, the vehicle is forced to travel fast short distances, zigzagging over the bottom with a rather erratic course. In this way, images can reflect the changes in texture and parallax that are necessary for a visual odometry. This, in turn, produces a stable estimation of the vehicle localization at the beginning of the exploration.

Once odometric data η and $\dot{\eta}$ are considered reliable, the adaptation of the controller coefficients is carried out by providing persistent and rich-in-frequencies changes in the reference position $\hat{\eta}_{ref}$ and rate $\hat{\dot{\eta}}_{ref}$ in the main degrees of freedom. To this end, it is convenient for $\hat{\eta}_{ref}$ to continue in time the previous erratic motion. Additionally, the rate $\hat{\dot{\eta}}_{ref}$ is defined as the sum of the cruising rate and a random sequence that causes soft accelerations and brakings along the trajectory. Once the convergence of the controller coefficients has been attained, the exploration starts at the current position. This position is marked as the starting point (shortened SP) and fixed for future trips.

Figure 2 (left) illustrates the construction of a corridor path from SP up to an end point (shortened EP), for instance, the estimated no-return point. The system analyzes the texture characteristics in the image sequence in order to detect robust clusters (C). Once they are recognized, the system tracks them as they pass in front of the camera. With a set of robust clusters, the system is able to triangulate frame by frame the pose of the camera.

Usually, raw information for robust cluster detection is provided by the feature-based SLAM method employed in the vision system. Certainly, for self-localization, the SLAM method constantly defines keyframes (KF) with the most stable keypoints in the image, i.e., with those that have been followed up so far. Thus, C-SLAM has to group the robust keypoints in clusters and finally perform an

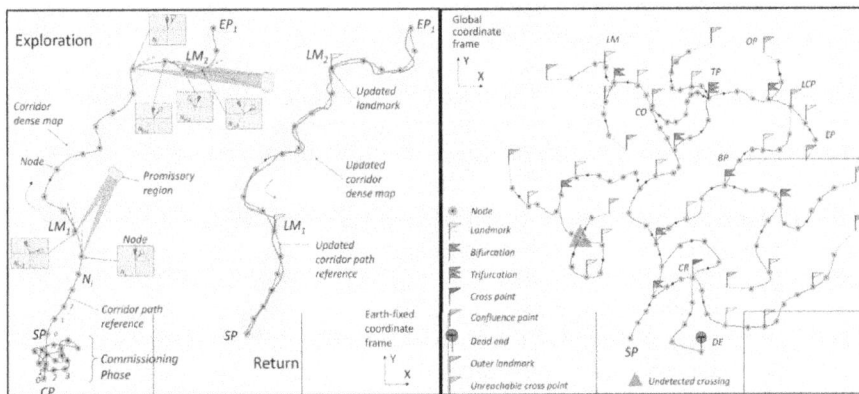

Figure 2.
Round trip during exploration (left). Corridor network topology (right).

evaluation of cluster set. In this way, C-SLAM can define nodes (N) and eventually bifurcation points (BP).

The topology of the corridor network usually contains a diversity of elements that are described in **Figure 2** (right), shortened SP, EP, KP, KF, C, N, LM, and BP. Additionally there are trifurcation points (TF) which are created from BPs: confluence points (CO) that occur between two paths that meet at a point; outer points (OP) which are detected landmarks or nodes but due to energy reasons are unreachable; dead-end points (DP) at which the vehicle must turn by force due to the absence of Ns or LMs in the horizon; and, finally, crosspoints (CR) at which a path crosses another path and these can or cannot be detected by the vision system.

It can frequently happen that, due to the existence of COs and CRs, many possible path alternatives are established to return from any point or to join two distant points in the network. Loops can also be generated during the expansion (see **Figure 2**, right).

As mentioned previously, the topological space is subject to an unknown and variable cartographic scale that produces accumulative odometric errors. This means that corridors can cross others without this being reflected in the global map. However, this does not invalidate the objective of C-SLAM, to return safely to SP.

It is expected that after reiterated revisits along different corridors, the number of loop closures carried out by the basic SLAM method will increase and the network will progressively take on a more real form. Simultaneously, the scale of the map will become more uniform, and the used metric will increasingly gain in accuracy.

The construction of the network is mainly based on a criterion to define a metric in the topological space. This is discussed below.

5. Nodes, landmarks, and keyframes

C-SLAM sets up a node to accurately synchronize future movement steps. For example, the direction to which the vehicle is to be driven in the short term in order to reach the next node is precisely defined at the current node. In addition, the speed of the vehicle on this direction is also defined here. All this gives the reference position $\hat{\eta}_{ref}$ and rate $\dot{\hat{\eta}}_{ref}$, which are used in the adaptive control system as inputs. Many other steps are synchronized at the current node as detailed below.

When C-SLAM decides to expand the network with a new branch, it searches for a landmark in the current corridor and then deflects the path in the indicated direction.

Nodes and landmarks are gestated in a common optimization process. Optimization involves maximization of a cost function with one optimum and many suboptimal results. The optimum outcome defines the first place in a ranking list, followed orderly by the suboptimal outcomes. Nodes and landmarks are selected from this ranking list.

The direction that defines $\hat{\eta}_{ref}$ and $\dot{\hat{\eta}}_{ref}$ to the next node is given by the optimal result. On the other hand, the best solution in the suboptimal set is reserved for a landmark.

Unlike a node whose next direction is defined in each step, a landmark can demand many steps. A landmark is a node whose high position on the merit list ranks it for a successful future branching. Therefore, not all the nodes become landmarks.

Finally, there is a subtle difference between a node and a keyframe generated from any SLAM technique. A node is the most robust cluster in the current keyframe, and in general its appearance frequency is generally lower than the frequency of a keyframe.

5.1 Selection of optimal directions and landmarks

During exploration, keyframes are the basis for selecting nodes. For every few keyframes of the sequence generated by the active SLAM algorithm, C-SLAM chooses the current keyframe and defines a node. The frequency of the node appearance depends physically on the visibility conditions and is commonly fixed in the commissioning phase.

On the other side, the selection of a path direction from the current node to navigate to the next (yet not defined) node follows a particular criterion. Also the selection of a promissory landmark for a future bifurcation is subject to the same criterion.

It is assumed that many keypoint clusters are faced at the same time in the ranking process in order to establish an order of merit. When the score of a particular cluster exceeds a threshold value, it is then considered as a potential landmark. Clearly, many clusters remain in the vision field for a time, so that they may continuously be tracked during their latency periods. Some of them will become landmarks.

After a landmark is lost in the field of vision, it will remain inactive with its last score and its location on the map. While in the exploration phase landmarks are generated, in a revisiting period they are consolidated, or even separated. Eventually, a landmark becomes a bifurcation point when C-SLAM decides to branch out to explore new regions.

The proposed criterion is based on a cost that is defined as a linear combination of quantifiable requirements to be met by each cluster in its latency period. Thus

$$V_i = \lambda_1 f_C + \lambda_2 T_C + \lambda_3 \Delta\theta + \lambda_4 \delta^{-1} \qquad (1)$$

where V_i is a cost for one keypoint cluster named C_i, given the weights λ_j with $0 \leq \lambda_j \leq 1$ and $\Sigma_j \lambda_j = 1$. The requirements for C_i are described and symbolized as

- High density of keypoints in a cluster: factor f_C

- Continuous traceability of a cluster: period T_C

- Robustness against the change of points of view: span angle $\Delta\theta$

- Alignment of a cluster on the heading direction: factor δ^{-1}

The cost weights will be defined previously according to the emphasis placed on some particular aspects of robust navigation. Since all requirement variables will be normalized to their expected highest values, a trivial choice $\lambda_j = 0.25$ is generally satisfactory. The simple election $\lambda_1 = 1$ should support a continuous navigation; however in this case zigzagging on the pathway should not be ruled out.

Since some detected landmarks are far away from the vehicle position, one should check whether these will be attainable with the current autonomy range of the vehicle. If they are unreachable, they will be removed from the ranking list or marked as OPs. To this goal a metric is required for the topological space of the corridor network.

5.2 Cost parameters

5.2.1 Density of keypoints in the cluster f_C

The key assumption to calculate the first parameter f_C is that visual features are very volatile in harsh underwater environments. Therefore, dense clusters may be thought to be more resistant to long-term disappearance than dispersed clusters. The first step to determine f_C is to group keypoints of the current keyframe in clusters. For this, for example, k-means clustering can be used in combination with an efficient initialization method. For the time k of cluster appearance, the following is calculated:

$$f_C(k) = \frac{N_{KP}(C_i)}{\max N_{KP}} \tag{2}$$

where i identifies the cluster of the set and N_{KP} is the number of KPs in C_i. Thus, the parameter can be determined recursively by

$$f_C = \frac{(k-1)f_C(k-1) + f_C(k)}{k} \tag{3}$$

Since the set of selected robust keypoints in the SLAM algorithm is generally sparse, clustering operation should not be excessively time-consuming.

Statistically, the maximum expected number of N_{KP} can be estimated in the commissioning phase as a function of the harshness of underwater environments. For example, poor visibility, self-similarity of the ground, caustic sunlight, shadows, flares, etc. are factors that contribute to reduce the number of robust keypoints.

5.2.2 Period for continuous traceability T_C

The second parameter is related to the traceability of C_i. The longer C_i is detected, the more feasible it will be to achieve continuity of navigation and a safe return. The parameter is estimated in each keyframe as follows:

$$T_C = \frac{N_{KF}(C_i)}{\max N_{KF}} \tag{4}$$

where N_{KF} is the number of consecutive KFs in which C_i appears approximately at the same position and almost with a similar density. To assess the similarity of the N_{KF} clusters, a Euclidean norm can be used. In addition, attention should be paid to changes in f_C and cluster midpoint. N_{KF} is estimated by thresholding all these variables properly.

Finally, the maximum number of consecutive keyframes depends to some extent on vehicle speed, altitude, camera tilt, and field of vision, among other factors. As a reference for $\max N_{KF}$ one can take the number of KFs required to cover a distance of 10 times the length of visibility in the line of sight.

5.2.3 Span angle $\Delta\theta$ for viewpoint range

The third parameter $\Delta\theta$ relates the robustness of C_i with the variation of viewpoints. A fairly large angle would be necessary for a cluster to be unambiguously detected from different points of view. $\Delta\theta$ applies in these two important cases: first when clusters are aligned in front of the vehicle and second when they are aligned to one side of the vehicle (see example in **Figure 2**, left, for landmarks LM1 and LM2). Thus, it results in

$$\Delta\theta = \frac{\text{span}\;\{\theta_j(C_i)\}}{\text{max span}\{\theta_j\}} \tag{5}$$

where the symbol span{.} means the difference between the maximum and minimum value of the sequence θ_j during the tracking of C_i. In turn, θ_j is the angle between the line that joins the camera with the cluster midpoint and the camera optical axis.

A large value of $\Delta\theta$ suggests a more robust link between the nodes and, on the other hand, in the case of landmarks, a more successful bifurcation.

The maximal span of θ_j is set equal to half of the horizontal field of vision of the camera, which specifies the tracking of a cluster that is initially seen far away in front of the vehicle and disappears from the field of vision sideways.

5.2.4 Course alienation factor δ^{-1}

The fourth parameter δ^{-1} plays a key role in the construction of the corridor. When the corridor is built step by step, one would normally prefer to keep the real course along a straight line rather than change the course. This is because the control of the vehicle along a line is generally more precise than on curves. Similarly, revisits on straight stretches are more immune to loop-closing failures than on curved stretches.

Therefore, in the case of nodes, δ^{-1} should generally produce a higher cost than in the case of landmarks that are reserved for bifurcations. Hence, landmarks are generally created on one side of the path, while nodes are rather in a straight line.

Since the cost (Eq. (1)) contains terms that must be maximized, δ^{-1} reflects the opposite of the alienation parameter here called β. Therefore, δ^{-1} should indicate how much the transverse separation of a cluster from the route is δ^{-1} assessed in the horizontal navigation plane. To this end, both the optical axis and the line that joins the camera with the midpoint of the cluster are projected onto the navigation plane. Between these lines, the angle β_j is obtained. A third line is defined, which connects the camera with the point furthest to the right of the vision field. Projecting this line on the navigation plane results in the angle called β^* with the projected optical axis.

Hence, it results in geometric mean

$$\delta^{-1} = \frac{1}{N}\sqrt{\sum_j \left(\frac{\tan\;\beta^*}{\tan\;\beta_j}\right)^2} \tag{6}$$

where N is the number of assessments j during cluster tracking.

5.3 Application of the selective criterion

The inclusion of a cluster C_i in the merit order list during its tracking is subject to

$$V_i > V^* \tag{7}$$

where V^* is a global threshold given for the cluster score. The merit order list is updated keyframe by keyframe. Hence, if C_i is already in the list, its score is simply actualized.

If V^* is set too high, many landmarks might be created, and the network could be unnecessarily too dense. The optimal number of bifurcations per corridor should satisfy a general rule that the distance between two consecutive BPs is spanned by about 20 nodes. On the other hand, if the branches are too scattered, V^* could be too high.

When a new node is created, C-SLAM immediately evaluates the next most promising direction to lead the vehicle into the unknown environment with the goal of exploration. Therefore, the highest score refers to the cluster with which the reference path $\hat{\eta}_{ref}$ should be continued.

In addition, C-SLAM checks whether the second highest score corresponds to an active cluster and in which case decides to set a landmark. Another more cautious strategy is to set a landmark when the second highest cluster just disappears from the list.

Landmarks can be also consolidated in the return and in every passage on any explored corridor. On the other side, the landmark cost may decrease so much that it is to be removed from the list. In this way, landmarks are permanently evaluated up to the moment they are employed to bifurcate into the corridor. At this moment, they change their status from LM to BP and disappear from the list forever.

In a changing environment, the adaptation of Ns, LMs, and BPs after a long period is necessary. This allows a renewal of the corridor network as needed (cf. [7, 30]).

During assiduous navigation, the topology of the network is outlined like a tree of branches. Each branch contains its identification in the tree, the sequence of nodes arranged in one direction, the landmarks, the bifurcation points, and any other particular element that should be important for decision-making.

6. Expansion of the network

Extending the boundaries of the corridor network requires C-SLAM to implement certain policy-makings, which goes beyond the corridor network building.

A first policy implies the definition of an appropriate metric to extend the lengths of the corridor network as much as possible according to the available energy autonomy.

A second policy supports the decision-making to choose bifurcations in order to multiply the number of branches. In this case, revisitings are of secondary importance as they take place only when they are needed as bridges to create new branches.

A third and final policy concerns the optimal scheduling of paths between two points of the network. In a situation with many alternatives to get a connection between sites, the system will search for the one with the best energy efficiency.

6.1 Metric

The metric space represented by the corridor network depends on an unknown scale which also changes over time due to cumulative odometric errors. This will affect, above all, the global map, which is the composition of many local maps with their own scale. As scale variations are generally small from map to map, the metrics are similar.

In order to maximize the length of a corridor without compromising the safety of the way back, it is essential to have a reliable metric. Since safety and energy autonomy are closely related, the metric must express the energy margin that the vehicle possesses at any time and position in the network (cf. [24, 25]).

It is clear that due to the challenging environment, the energy used to connect two points is often not the same in both directions. This will undoubtedly depend on sea currents, vehicle speed changes, travel breaks, and zigzagging on the reference route. Thus, for example, real-time detection of the no-return points based on the battery state of charge may not be reliable enough in unforeseeable circumstances. For this reason, a more powerful detection of no-return points is developed here on an empirical basis.

To this end, a function for the estimation of the energy margin is proposed. It adds up the energy consumption until the full autonomy is completed. The function can be evaluated at any time, especially in the decision to return.

The idea behind the proposal is that almost all the energy available in navigation is intended to move the vehicle. Therefore, an approach that is based solely on the energy of motion along the travelled path seems to be quite rational.

The first one defines the consumption of energy as a set of possible cases:

$$E\left(\hat{\dot{\eta}}, \hat{\eta}\right) = \left\{ \int_{C_i} \hat{\dot{\eta}}^2(\hat{\eta}) d\hat{\eta} \right\} \tag{8}$$

where $\{.\}$ is a symbol for a set of cases and $\int_{C_i} \_ d\hat{\eta}$ describes a curvilinear integral along the path i that spans the way from the SP up to the point where the autonomy is completed (see **Figure 3** to support the concept). Similarly, the distance covered by the vehicle on this path i until all available energy is consumed is defined as a set of cases:

$$d(\hat{\eta}) = \left\{ \int_{C_i} d\hat{\eta} \right\} \tag{9}$$

It should be noted that all routes cover each shape, both straight and curved ones.

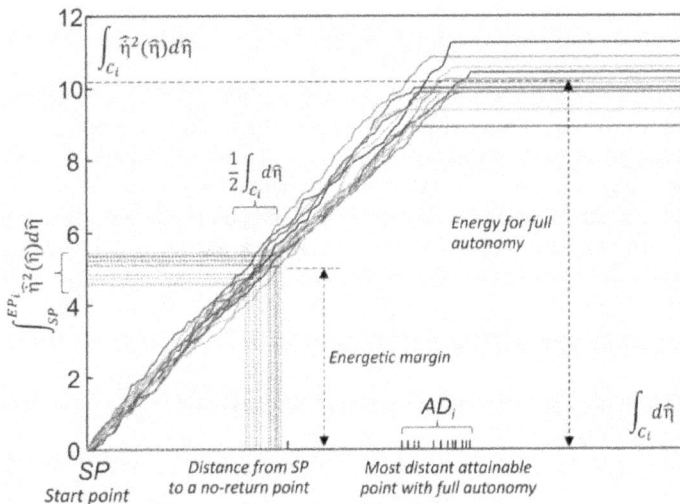

Figure 3.
Energy margin to reach total energy consumption. Numerical identification of no-return points on different paths.

Figure 3 shows numerical simulation results wherein the distance travelled by a vehicle over different paths is computed after accomplishing full autonomy. In the experiments the vehicle acquires a cruise velocity, but this is slightly changed around this value at random. The resulting ground truth statistics must be available before applying C-SLAM but are updated during the construction of the corridor network.

In order for C-SLAM to decide the time point for the comeback on a new corridor, a norm based on the average on the sets in Eqs. (8) and (9) is applied. The Euclidean norm can be used to obtain the energy limits called E_{EP} and travelled distance d_{EP}:

$$\|E_{EP}\|_2 = \frac{1}{N}\sqrt{\sum_i \left(\int_{SP}^{EP_i} \hat{\eta}^2(\hat{\eta})d\hat{\eta}\right)^2} \quad and \quad \|d_{EP}\|_2 = \frac{1}{2N}\sqrt{\sum_i \left(\int_{C_i} d\hat{\eta}\right)^2} \tag{10}$$

subject to the conditions for safe navigating on the path i

$$\|d_{EP}\|_2 \geq \int_{C_i} d\hat{\eta} \tag{11}$$

$$\|E_{EP}\|_2 \geq \int_{C_i} \hat{\eta}^2(\hat{\eta})d\hat{\eta} \tag{12}$$

where the equality of either of the two equations will mean the identification of the end point EPi for this path. N is the number of trips in the statistics that cover a distance d_i for any route from SP. It is very important to frame the confidence of the norm in the risk of loss of the vehicle. In this sense, the most unfavorable case might be more appropriated. In this case, it is valid

$$\|E_{EP}\|_{max} = \frac{max}{C_i}\left\{\int_{SP}^{EP_i}\hat{\eta}^2(\hat{\eta})d\hat{\eta}\right\} \quad and \quad \|d_{EP}\|_{min} = \frac{min}{C_i}\left\{\frac{1}{2}\int_{C_i}d\hat{\eta}\right\} \tag{13}$$

and so the conditions for the detection of a no-return point are as follows:

$$\|d_{EP}\|_{min} \geq \int_{C_i} d\hat{\eta} \tag{14}$$

$$\|E_{EP}\|_{max} \geq \int_{C_i} \hat{\eta}^2(\hat{\eta})d\hat{\eta} \tag{15}$$

where again the equality of either of the two equations will mean the identification of the end point EPi for this path i.

In order to continuously adapt the norm to the environment, the set (Eqs. (8) and (9)) must be updated on each round trip. It can be noticed that the update can be applied both in a new corridor and in a revisiting case as well, regardless of the path chosen.

6.2 Planning of new corridors

The way in which C-SLAM progressively explores and increases branches can be supported by different criteria.

6.2.1 Method I

The first method is based on the criterion that for fast expansion of new branches, occurrences of revisitings of old branches should be reduced as much as possible. To this end, the number of revisited stretches in old branches and their distances to SP

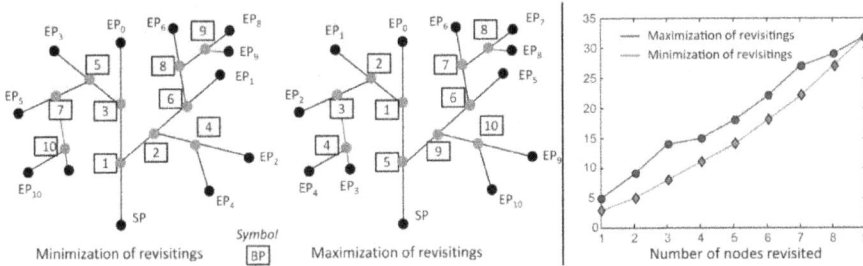

Figure 4.
Generation of BPs according to different criteria: the fastest branch expansion (left) and the slowest branch expansion (middle). Number of revisited stretches during network expansion: optimal expansion (red) and a suboptimal expansion (blue).

should be minimized during the sequence of round trips. Here, the rectification of the path $\int_{SP}^{BP_i} d\hat{\eta}$ is employed. It is evident that to reach a certain BP on an old branch, the vehicle must unfailingly revisit some first stretches on this branch.

Figure 4, right and middle, illustrates a case study of an expanded network in two different forms.

Once the trunk corridor is created, the proposed method starts from SP and searches for the closest LM and splits into a new branch until it culminates in an EP. At this stage, this LM becomes a BP. In doing so, the approach minimizes the length of all revisited stretches. Any other branch generation is suboptimal. The worst suboptimal generation, on the contrary, expands the network by starting each round trip from SP to the most distant LM. In fact, this option produces a maximization of the revisited stretches.

Even when the optimal branch generation expands the network faster than other options, all methods end up with the same number of revisited stretches. This fact is reflected in **Figure 4** (right). Therefore, from the point of view of the total energy consumed, the difference is short and medium term only.

6.2.2 Method II

The optimal solution described above has a theoretical rather than a practical value. The disadvantage of method I is that the way BPs are used has no connection with the merit order list. For that reason, vehicle safety might be compromised. When making decisions about where to branch, trust in the list is the only support for secure expansion.

Therefore, the main focus of the second method is to build up a solid network in which the path from every EPi to SP maximizes the trust on the BPs on the pathway. This means that the LMs chosen for bifurcation have the maximal score on this pathway. The number of revisited stretches over time is described by a curve that lies between the two extreme curves in **Figure 4** (right). This means that in the long term, the new curve will also converge to the same extreme value.

Many other strategies can be applied besides the other two methods. For instance, one could be specifically interested in developing the branches to the left (or to the right) of the trunk line or in creating a solid statistic for the newly explored branch before continuing with the exploration.

6.3 Path scheduling

Starting from the conception of a fully interconnected corridor network, multiple alternatives can be considered to obtain a connection between two distant

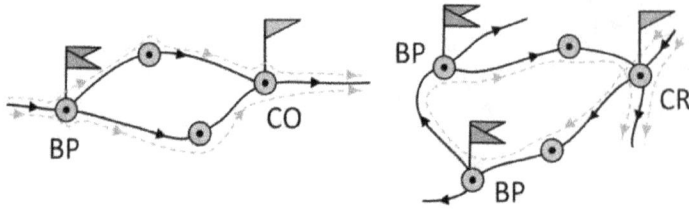

Figure 5.
Bypass (left) and loop (right) in the network topology.

points of the network (see, for example, in **Figure 3**, right, the bypass shaped by the points BP-CO and BP-TP-CO). In such situations, C-SLAM should be able to seek that path that involves the best energy efficiency. For that purpose C-SLAM must count with odometric information of each stretch and the energy consumption cost per unit of length.

Once the network has converged in its expansion and especially after a large number of revisits, it is common to count with the presence of bypasses and loops. Thereupon these elements will become part of the topology of the network.

In the case of one-way corridors, the existence of a bypass can only occur in the presence of COs or CRs (see **Figure 5**, left). On the other hand, the generation of a loop needs indefectibly at least one CR (see **Figure 5**, right). It is worth noticing that a loop represents a dummy alternative to connect a point with oneself through the loop as illustrated in **Figure 5** (right).

As topological elements of the network, loops are marked to avoid unnecessary energy consumption. For example, in **Figure 5** (right) when a loop is encountered, the path entering the loop is avoided unless the destination is a BP allowing exit from the loop. For the optimization of paths, the criterion consists in searching for the shortest path between two points or for the path that minimizes the energy consumption, or a combination of both. If safety aspects are emphasized, the energy approach is the most suitable for C-SLAM. This option is described in the following.

To find the optimal path, the following cost functional is minimized for a path starting at node N_A and ending at node N_B:

$$J(N_A, N_B) = \min_{\{N_A, N_B; C_k\}} \left\{ \sum_{N_A}^{N_B} \int_{N_i}^{N_{i+1}} \hat{\ddot{\eta}}^2(\hat{\eta}) d\hat{\eta} \right\} \tag{16}$$

where C_k is one of the paths that connects N_A and N_B and the pair N_i and N_{i+1} is the terminal nodes of the stretch in C_k. The energy information on the stretch from N_i to N_{i+1} is the result of an average each time that the vehicle passes on this section. For the minimization in Eq. (16), BPs and COs, even dead-end points, are taken into account along with all nodes. Thus, the minimal value of the sum on every stretch that links N_i and N_{i+1} of C_k represents the optimal path denoted by C_k^*.

To assist Eq. (16), the A^* search algorithm has been chosen in this work over other tree search algorithms because of its optimal efficiency [31]. Therefore, the cost is

$$F(N_{i+1}) = g(N_{i+1}) + h(N_{i+1}) \tag{17}$$

where g is the cost of the path from N_A to N_{i+1} and h is a heuristic function that estimates the cost of the cheapest path C_k^* from N_{i+1} to N_B. A* ends when the goal is reached or when there are no paths eligible to be extended. In this work, the cost functions g and h are selected according to the following two terms:

$$F(N_{i+1}) = J(N_A, N_{i+1}) + \frac{min}{\{N_{i+1}, N_B; C_k\}} \left\{ \sum_{N_{i+1}}^{N_B} \int_{N_j}^{N_{j+1}} \hat{\eta}^2(\hat{\eta}) d\hat{\eta} \right\} \qquad (18)$$

The heuristic function $h(N_{i+1})$ is admissible provided that paths with one or more sections travelled in both directions are excluded from the minimization. Thus, branches that do not represent a bypass to the branch on which N_B is located are discarded. Loops are also avoided when used to connect a node to itself.

7. Case study

This section shows some experimental results which are selected to illustrate the viability of the proposal for autonomous navigation based on the proposal. Many of the functions of the approach are tested together according to the C-SLAM structure in **Figure 1**. It is important to note that the case studies are carried out in limited spaces with simulation of light effects and seabed textures. However, the staged environments reflect quite well the difficulties encountered in image processing in real underwater scenarios.

7.1 Environment

In order to provide a ground truth for testing and to achieve acceptable reproducibility of results, many scenarios were set up in a basin with a staged underwater landscape that closely resembled the natural seabed. In this scenario, rocks, gravel, sandbanks, and benthos, among others, predominated wherein a variety of underwater visual effects could be obtained [19, 23] (see **Figure 6**).

The bioactivity of microorganisms was cyclically changing the characteristics of soil texture and water transparency. The scenarios were illuminated by direct and indirect sunlight. However, light disturbances and turbidity were controlled for the

Figure 6.
Stage installations for tests underwater. Diversity of scenarios.

range of tests. For instance, strong or weak sunlight flickers on the ground were generated by agitating the water in two orthogonal directions. In addition, visibility was reduced by discharging silt particles onto the surface that remained suspended for a period of time.

Therefore, a wide variety of case studies could be faithfully reproduced, such as rapid changes in luminance, transition from dark to bright scenarios, blurriness, lens flares, motion blur, self-similarities, glare, and bubbles (see **Figure 6**).

7.2 Hardware

An ROV (model OpenROV v2.8) was used as the platform for the experiments, although this was hydrodynamically reformed with side fins to reduce the motion blur.

Two independent cameras were installed on board, a high-resolution, wide-angle vehicle camera (Genius f100) and a high-performance camera (GoPro Session H4). Both cameras are rolling shutter and operate at different frame rates. To attenuate undesired effects of the rolling shutter mechanism, especially in photometric-based algorithms, a high speed of 120 fps with an image size of 848 × 480 pixels was used for dense mapping, albeit off-line at the end of every round trip.

The vehicle was completely steered by C-SLAM that was programmed and installed in a notebook with GPGPU technology. The bi-directional flow of video and control signals was implemented via cable. The altitude was conveniently fixed in advance to adapt to the visibility of the environment. Altitude control was carried out independently of the adaptive controller by a PI controller. An adaptive speed-gradient controller [17] was used to self-adjust the controller coefficients for the main dynamics of the ROV.

7.3 Software

Among other functions, C-SLAM uses state-of-the-art free software to localization and mapping. As shown in **Figure 1**, the block for feature-based SLAM was implemented with the algorithm ORB-SLAM [32]. On the other hand, the block for dense mapping was implemented with the photometric-based DSO-SLAM [33] and sometimes with LSD-SLAM [34, 35]. However, there was a significant modification in the implementation, namely, DSO (or LSD-SLAM) was supported with the estimated camera position provided by ORB. This was necessary in general for improving the stability of the mapping in very harsh environments, especially due to the presence of strong sunlight caustics or/and poor visibility [23]. The deflickering filter in this work is based on estimations of sunlight caustic fringes using a feedback of predicted images [20], which employs very high accuracy velocity estimation [36]. Software for guidance, control, and corridor construction was developed specially for C-SLAM.

7.4 Results

One begins with **Figure 7** which illustrates the role deflickering filter plays in the improvement of dense mapping in environments with strong sunlight caustics. Generally, spatiotemporal light changes affect the performance of photometric-based methods seriously. As seen in the heat maps, the camera depths are coherent with the physical environment. Thus, the photometric consistence is preserved after image deflickering.

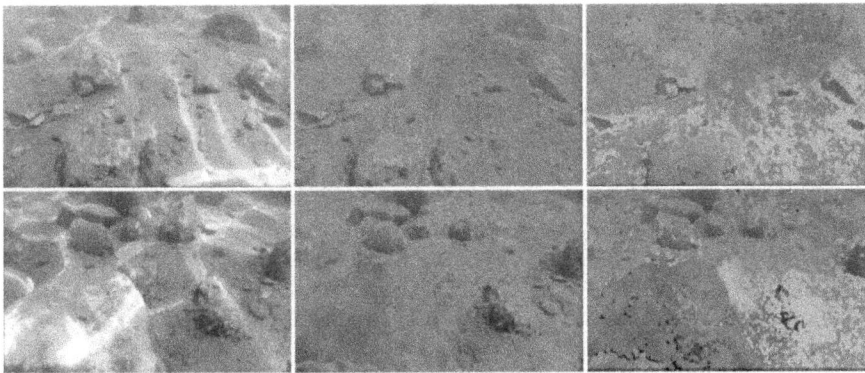

Figure 7.
Dense mapping in two cases after image deflickering. Images with sunlight caustics (left), deflickered images (center) [14], and heat maps for camera depth (right).

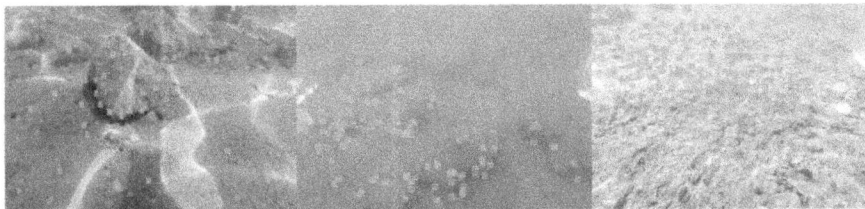

Crop bottom to Height=1,75 cm

Figure 8.
Selection of robust keypoints in three harsh environments: with strong sunlight caustics (left), low visibility (center), and self-similarities (right).

The situation is totally different in the case of ORB-SLAM, where light disturbances do not affect ORB to the same degree as DSO (or LSD-SLAM) (see **Figure 8**). However, from several experiments, it was concluded that in turbid waters, robust texture features decrease substantially, although the overall performance of ORB does not degrade as much as in the case of DSO (see [8]). Therefore, the filtering of caustic sunlight waves can be omitted in the case of ORB, but not in the case of DSO (or LSD-SLAM). For this reason, the deflickering filter in the C-SLAM structure in **Figure 1** is necessary only for dense mapping ends. Another conclusion was drawn from navigation in self-similar terrains staged in the basin like in the third picture to the right in **Figure 8**.

These terrains provide commonly numerous features in a similar cluster pattern, but ORB-SLAM often losses the track as it is unable to deal with nearby similarities.

An important instance at the beginning of C-SLAM navigation is the start-up phase for initializing SLAM algorithms and adaptive controller parameters. To this end, in the study, the vehicle movements were performed manually providing a zigzag path of the camera. **Figure 9** illustrates the initial process of constituting a dense mapping of the environment with an adequate camera trajectory. From there, the start of the exploration was supplied with good estimates of vehicle position and speed, which in turn, allow the adaptive controller to adjust its coefficients.

Dense methods suffer, more than any other class, from odometric errors [8], which are minimized in the particular case of DSO through a very cumbersome and thorough camera calibration process. In these trials, the combination of direct methods for mapping with ORB-SLAM for tracking increases the accuracy of the global map of the corridor network, even in the case of normal rolling shutter cameras as used here. **Figure 10** illustrates these results in one experiment with good textured floor.

The limited space and relatively short time span for the experiments prevented the use of an energy autonomy metric beyond the context of numerical computer simulations. In this sense, end points located sideways must be at a certain limited distance from the trunk corridor. The frequency of occurrence of nodes was synchronized with the keyframe generation according to the simple strategy, namely, "one keyframe, one node." The direction to the next node was optimized primarily by the maximizing the density of cluster keypoints near the front line of vision to avoid zigzagging of the vehicle. In this way, branches could be extended to a relatively significant length.

An important feature of the monocular C-SLAM is the vehicle return through the corridor in reverse motion as seen in the display provided by ORB-SLAM by means of the symbol "⬡⇨," indicating the direction of movement (see **Figure 11**, top). In addition, the track link that exists when the algorithm identifies a connection between nodes is characterized by green segments between them (see **Figure 12**, left).

Reverse movements often caused motion blur due to pulling and cable drag on the floor [37]. Besides, the drag of the ROV backside is much more pronounced than

Figure 9.
Commissioning phase to initialize photometric dense mappings. Camera path on dense map (left), camera depth (top-right), and original frame (bottom-right).

Figure 10.
Estimated camera path and dense map. Map and path (left), heat map of camera depth (top-right), and original frame (bottom-right).

in the forward displacement. All this made it necessary to reduce the cruise velocity to minimize large heading perturbations that would cause track loss. In the following, two case studies illustrate the C-SLAM performance under these circumstances.

Figure 11 shows a corridor network that was constructed under the criterion of fast expansion of new branches, i.e., minimizing of revisits in short and medium terms. Here, three bifurcation points were implemented in four steps. First, the system creates the trunk corridor and returns to SP. Accordingly, in the next round trip, the vehicle is led to the nearest BP and forced to bifurcate to create the first branch, and thereafter it returns to SP again. This routine was repeated for the second and third BP.

It is notorious that the way back through the trunk corridor from the BP3 to the SP does not completely agree with the way out. However the position track was never lost, which demonstrates the robustness of the system against changes of points of views. The differences in trajectories were due to many causes, including control tracking errors, cable tugs, and drag disturbances [37].

The diversity of terrain texture shown in **Figure 11** (bottom) and their impact in the performance and robustness of C-SLAM are noticeable. For instance, at BP_1, the terrain is bulky, so the contrast is high and the keypoints are robust. On the other hand, in BP2 and BP3, the terrain was acquiring an increasingly self-similar appearance, and the keypoint clusters were becoming increasingly volatile. In some trips in reverse through this self-similar zone, the vehicle briefly lost its tracked position.

Figure 12 illustrates a more sophisticated experiment in which branches were allowed to occur on both sides of the trunk corridor. Over time, in addition to the formation of branches, confluence of paths and way crossings were also appearing. The picture on the left shows the network from above and the interconnection of nodes. On the other side, the picture on the right displays the network nodes and

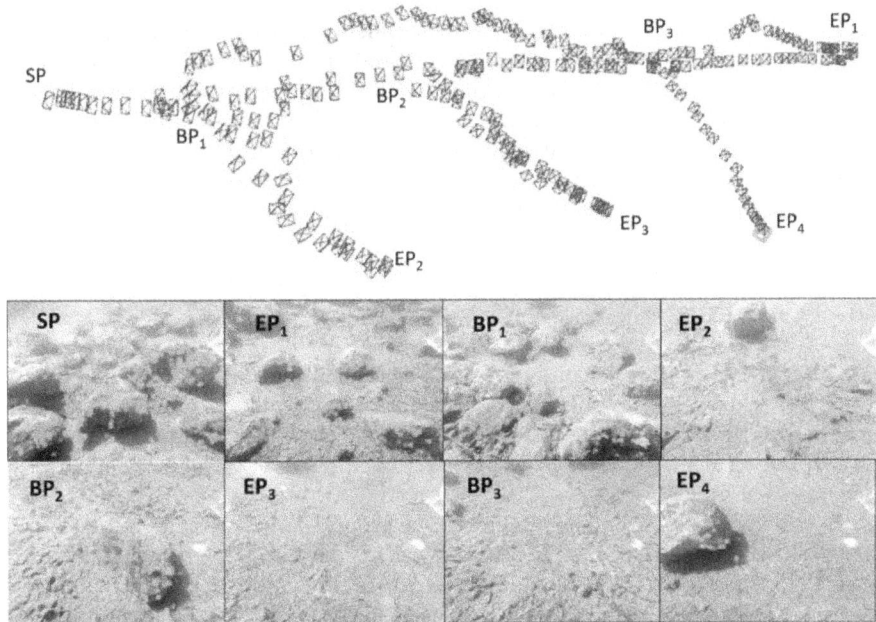

Figure 11.
Corridor network construction with three branches (top). Camera-taken scenes at the corresponding SP, BPs, and EPs (bottom).

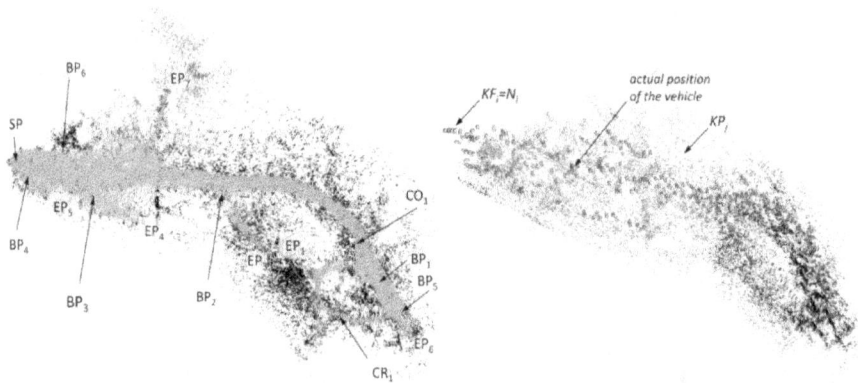

Figure 12.
Construction of a corridor network. Path graph with localization of BPs, EPs, COs, and CRs (left). Node sequence in a sparse map (right).

keypoints produced during the multiple round trips. It is observed that the density of keypoints is significantly higher in the right zone than in the left zone of the explored area. This is due to the marked self-similarity of the terrain that dispersed the keypoints over the terrain, while in the left zone, the bulky elements concentrate the keypoints around their peripheries.

The trunk corridor was long enough to encompass areas of different textures. Light disturbances were moderate; most of them transitions from dim to shining scenarios.

The sequence of the round trips can be followed by the order of occurrence of EPs and BPs. The strategy of expanding the network is the same as in the case before, i.e., create the trunk corridor, return to SP, annex a new branch, and begin a new round trip.

The most remarkable instance in this experiment occurred on a stretch in front of EP1, in the self-similar zone, where the C-SLAM briefly loses the vehicle location. Afterwards the vehicle was led in the direction of EP1 up to cross the trunk corridor. This CR was recognized by the system and aggregated to the network. From this point, the vehicle could return in reverse to a point that was also recognized by the system as a CO. Finally, after the sixth branch, C-SLAM completed the exploration by providing a global map of the corridor network.

8. Conclusions

This work deals with the autonomous navigation of underwater vehicles with the aim of achieving a broad exploration of the seabed. Unlike autonomous navigation systems in aerial, terrestrial, and space applications, for which the main source for localization is a GPS system, this approach basically uses only a monocular camera as sensor.

To establish the position of the vehicle, the proposed vision system takes advantage of the characteristics of texture of the seabed. Since underwater scenarios are generally very diverse and harsh due to water turbidity and light disturbances, vision-supported navigation poses a huge challenge for safe autonomous exploration. In particular, the lack of transparency in water forces the control system to lower the altitude to the seabed, which in turns demands a high degree of maneuverability to avoid collisions.

This work describes a vision-based system named C-SLAM that provides a nested loop structure to integrate control, guidance, navigation, and route planning systems into one. To explore the seabed, C-SLAM implements a strategy based on active SLAM. In contrast to many methods in the literature that employs topological models of the environment like feature graphs [13], Bayes tree [14], or grid maps [15], C-SLAM presents the environment as a network of interconnected corridors made up nodes and bifurcation points. It is claimed in this work that this simple topology is highly adequate and robust for harsh underwater environments with self-similarities, strong spatiotemporal light perturbations, and lack of water transparency [23].

Compared to other optimization techniques proposed for underwater applications such as random trees [38], particle swarming [36], octo-trees [39], level setting [40], and genetic background [41], among others, this proposal has a significant level of novelty. It radially searches in the field of vision for the optimal direction to explore. This direction is subject to satisfy quantifiable requirements for the navigation. The requirements are integrated in a weighted linear combination that is maximized step by step. This produces an active scoring list of promising points for selection of nodes and future bifurcation points providing adaptation to the environment and robustness of the node sequence.

The heuristic of the proposal have similitudes with the visual teach and repeat method for long-range autonomy [16]. Therein a pathway is created in an unfamiliar outdoor environment, at first by teleoperation and then followed by a rover. However, C-SLAM is essentially designed for autonomous exploration underwater and deals with a dense exploration in 2D. Besides, the submarine navigates basically on a plane over the seabed contemplating topographic features from above, not over the terrain. So, comparatively, C-SLAM has the advantage to choose the optimal route in order to be able to safely return to the start point. Another difference is that the C-SLAM can cope with the lack of map scale and odometric errors and even so ensure the vehicle return.

In order to expand the explored region as far as possible, a suitable metric in the non-scaled topological space is defined in relation to the vehicle energy autonomy. The key idea is to lead the vehicle by a corridor just up to the no-return point in order to make the return trip safe. In contrast to other approaches that solve the problem in a rather complex form, for instance, employing dynamic programming onto bathymetric and sea current maps [26, 27], C-SLAM proposes a statistic-based connection of odometric information and energy consumption.

The approach was experimentally tested in reduced-scale in a basin, wherein subaquatic environments with a good resemblance to a real seabed were staged. Experimental trials have demonstrated the feasibility of the approach in future applications where an autonomous underwater vehicle can host the C-SLAM vision system for large-scale underwater exploration. With the exception of extremely turbid environments, practically in all other cases, C-SLAM was able to make correct decisions to create and expand the underwater corridor network in a stable manner.

Acknowledgements

I thank E. Trabes for the valuable assistance in implementing the concepts and ideas elaborated in this work. I also thank J.L. Bustamante, J.L. Bonitatibus, G.D. Van Waarde, and L. Nuciari for their support in the experiments carried out at IADO. This work was financially supported by CONICET, CCT-Bahía Blanca, Argentine.

Author details

Mario Alberto Jordan
Argentinean Institute of Oceanography (IADO-CONICET), Electrical Engineering
and Computer Department (DIEC), University of the South (UNS), Bahia Blanca,
Argentine

*Address all correspondence to: mjordan@criba.edu.ar

IntechOpen

References

[1] Lowry S, Snderhauf N, Newman P, Leonard JJ, Cox D, Corke P, et al. Visual place recognition: A survey. IEEE Transactions on Robotics. 2016;**32**(1):1-19

[2] Cadena C, Carlone L, Carrillo H, Latif Y, Scaramuzza D, Neira N, et al. Past, present, and future of simultaneous localization and mapping. Towards the robust-perception age. Open access arXiv.1606.05830v2. 2016

[3] Horgan H, Toal D. Computer vision applications in the navigation of unmanned underwater vehicles, Chap. 11. In: Inzartsev AV, editor. Robotics, Mobile Robotics, Underwater Vehicles. 2009. DOI: 10.5772/6703. ISBN 978-953-7619-49-7

[4] Sonka M, Hlavac V, Boyle R. Image Processing, Analysis, and Machine Vision. Stamford (Connecticut): Thomson; 2008. ISBN 0-495-08252-X

[5] Agarwal S, Lazarus SB, Savvaris A. Monocular vision based navigation and localization in indoor environments. In: Proceedings of IFAC Workshop on Embedded Guidance, Navigation and Control in Aerospace, Bangalore, India. Vol. 45(1). 2012

[6] Li J, Arbor A, Eustice RM, Johnson-Roberson M. Underwater robot visual place recognition in the presence of dramatic appearance change. In: Proceedings of the IEEE OCEANS 2015 —MTS/IEEE Washington DC; IEEE. 2015. pp. 1-6

[7] Smith RN, Py F, Rajan K, Sukhatme GS. Adaptive path planning for tracking ocean fronts with an autonomous underwater vehicle. In: The 14th International Symposium on Experimental Robotics. 2014. pp. 761-775

[8] Oleari F, Kallasi F, Rizzini DL, Aleotti J, Caselli S. An underwater stereo vision system: From design to deployment and dataset acquisition. In: Proceedings of the OCEANS 2015; 18-21 May; Genova. IEEE; 2015, p. 1-6

[9] Lee D, Kim D, Lee S, Myung H, Choi HT. Experiments on localization of an AUV using graph-based SLAM. In: Proceedings of the 10th International Conference on Ubiquitous Robots and Ambient Intelligence (URAI). IEEE; 2013. pp. 526-527

[10] Folkesson J, Lonard J. Autonomy through SLAM for an underwater robot. In: Robotics Research. Springer Tracts in Advance Robotics; 2011

[11] Hessner K, Reichert K, Rosenthal W. Mapping of sea bottom topography in shallow seas by using a nautical radar. In: 2nd International Symposium on Operationalization of Remote Sensing; 16-20 August; Enschede, The Netherlands. 1999

[12] Trabes E, Jordan MA. Node-based method for SLAM navigation in self-similar underwater environments: A case study. MDPI Robotics. 2017; **6**(4):29

[13] Chaves SM, Eustice RE. Efficient planning with the Bayes tree for active SLAM. In: International Conference on Intelligent Robots and Systems (IROS), 2016 IEEE/RSJ; Daejeon, South Korea. IEEE; 2016

[14] Maurovi I, Seder M, Lenac K, Petrovi I. Path planning for active SLAM based on the D* algorithm with negative edge weights. In: IEEE Transactions on Systems, Man, and Cybernetics: Systems. Piscataway, New Jersey: IEEE; 2017. p. 99

[15] Mu B, Giamou M, Paull L, Agha-Mohammadi AA, Leonard J, How J. Information-based active SLAM via

topological feature graphs. In: IEEE 55th Conference on Decision and Control (CDC). IEEE; 2016

[16] Furgale P, Barfoot TD. Visual teach and repeat for long-range rover autonomy. Journal of Field Robotics. 2010;27(5):534-560

[17] Jordán MA, Bustamante JL. Adaptive control for guidance of underwater vehicles. In: Inzarsev A, editor. Underwater Vehicles. Vienna, Austria: In-Tech; 2009. pp. 251-278. ISBN: 978-953-7619-49-7

[18] Conor M, Py F, Rajan K, Ryan J. Adaptive control for autonomous underwater vehicles. In: Proceedings of the Twenty-Third AAAI Conference on Artificial Intelligence, AAAI 2008; Chicago, Illinois, USA; July 13-17. p. 2008

[19] Swirski Y, Schechner YY, Herzberg B, Negahdaripour S. Underwater stereo using natural flickering illumination. In: IEEE OCEANS'10, Sydney. 2010

[20] Trabes E, Jordan MA. On-line filtering of sunlight caustic waves in underwater scenes in motion. In: 7th International Scientific Conference on Physics and Control; 19-22 August 2015; Istanbul, Turkey. p. 2015

[21] Roosmalen PMBV, Lagendijk RL, Biemond J. Flicker reduction in old film sequences. In: Time-varying Image Processing and Moving Object Recognition 4. Amsterdam, Netherlands: Elsevier Science; 1997. pp. 9-17

[22] Gracias N, Negahdaripour N, Neumann L, Prados R, García R. A motion compensated filtering approach to remove sunlight flicker in shallow water images. In: Proc. MTS/IEEE Oceans. 2008

[23] Jordan M, Trabes E, Delrieux C. A robust approach for monocular visual odometry in underwater environments.

In: Proceedings of the ASAI, 48 JAIIO. 2019

[24] Chellappa R, Qian G, Srinivasan S. Structure from motion: Sparse versus dense correspondence methods. In: Proc. Int. Conf. on Image Processing, 1999. ICIP 99. 1999

[25] Jordan M, Bustamante JL. On the autonomy in unmanned underwater vehicles with the most distant point of no return. In: VI Jornadas Argentinas de Robótica; 3 al 5 November; Buenos Aires, Argentina. 2010

[26] Chyba M, Haberkorn T, Singh SB, Smith RN, Choi SK. Increasing underwater vehicle autonomy by reducing energy consumption. Ocean Engineering. 2009;36:62-73

[27] Zeng Z, Lian L, Sammut K, He F, Tang Y, Lammas A. A survey on path planning for persistent autonomy of autonomous underwater vehicles. Ocean Engineering. 29 June 2019;9 (2654):1-22

[28] Low KH, Dolan JM, Khosla P. Adaptive multi-robot wide-area exploration and mapping. In: Proceedings of the 7th International Joint Conference on Autonomous Agents and Multiagent Systems. Vol. 1. 2008. pp. 23-30

[29] Zhu D, Huang H, Yang SX. Dynamic task assignment and path planning of multi-AUV system based on an improved self-organizing map and velocity synthesis method in three-dimensional underwater workspace. IEEE Transactions on Cybernetics. 2013; 43(2):504-514

[30] Alvarez A, Caiti A, Onken R. Evolutionary path planning for autonomous underwater vehicles in a variable ocean. IEEE Journal of Oceanic Engineering. 2004;29(2):418-429

[31] Guruji AK, Agarwal A, Parsediya DK. Time-efficient A*

algorithm for robot path planning. Procedia Technolgy. 2016;**23**: 144-149. DOI: 10.1016/j.protcy.2016. 03.010

[32] Mur-Artal R, Montiel JMM, Tardós JD. ORB-SLAM: A Versatile and Accurate Monocular SLAM System. Piscataway, New Jersey: IEEE; 2015. DOI: 10.1109/TRO.2015.2463671

[33] Engel J, Koltun V, Cremers D. Direct sparse odometry. IEEE Transactions on Pattern Analysis and Machine Intelligence. 2018;**40**(3):611-625

[34] Engel J, Schöps J, Cremers D. LSD-SLAM: Large-scale direct monocular SLAM. In: Proc. of the European Conf. on Computer Vision (ECCV). Midtown Manhattan, Nueva York: Springer; 2014. pp. 834-849

[35] Newcombe RA, Lovegrove SJ, Davison AJ. DTAM: Dense tracking and mapping in real-time. In: Proceedings of the IEEE International Conference on Computer Vision (ICCV). Barcelona, Spain: IEEE; 2011. pp. 2320-2327

[36] Farnebäck G. Very high accuracy velocity estimation using orientation tensors, parametric motion, and simultaneous segmentation of the motion field. In: Proc. Eighth International Conference on Computer Vision. Vol. 1. Vancouver, Canada: IEEE Computer Society Press; 2001. p. 171-177

[37] Jordan M, Bustamante JL. Numerical stability analysis and control of umbilical-rov systems in taut-slack condition. Nonlinear Dynamics. 2006; **49**:163-191

[38] Rao D, Williams SB. Large-scale path planning for underwater gliders in ocean currents. In: Australasian Conference on Robotics and Automation (ACRA); December 2-4, 2009; Sydney, Australia. Vol. 110, Part A. Amsterdam, Netherlands: Elsevier; 2015. pp. 303-313

[39] Yan ZA, Li J, Wu Y, Zhang G. Real-time path planning algorithm for AUV in unknown underwater environment based on combining PSO and waypoint guidance. MDPI. 2018;**19**(1). Article number: 20

[40] Hernández JD, Vidal E, Vallicrosa G, Galceran E, Carreras M. Online path planning for autonomous underwater vehicles in unknown environments. In: Proc. IEEE Int. Conf. on Robotics and Automation. 2015. DOI: 10.1109/ICRA.2015.7139336

[41] Liang S, Zhi-Ming Q, Heng L. A survey on route planning methods of AUV considering influence of ocean currents. In: Conference: 2018 IEEE 4th Int. Conf. on Control Science and Systems Engineering (ICCSSE); 21-23 Aug. 2018; Wuhan, China

Underwater Technical Inspections Using ROV Applied to Maritime and Coastal Engineering: The Study Case of Canary Islands

Sérgio António Neves Lousada, Rafael Freitas Camacho and Josué Suárez Palacios

Abstract

Underwater Technical Inspections using ROV have an important role in the design, construction, maintenance and repair of maritime and coastal infrastructures, trough video recording, digital photographs, collection of technical data and underwater topographic survey providing support for consultancy studies and projects and technical advice and appraisals. Routine inspections are the key to the maintenance of any submerged infrastructure. The importance of this type of inspection is increasing every day, but divers are also placed in increasingly dangerous scenarios to carry out this type of work. Inspections of underwater structures (as in dams, bridges, reservoirs, breakwaters, piers, oil rigs, etc.) have always been arduous and difficult, and often dangerous, but today underwater drones offer solutions that eliminate the risk faced by divers, and that also greatly reduce the high costs involved in such inspections.

Keywords: coastal engineering, construction, data, design, maintenance, maritime engineering, ROV, supervision, underwater inspections, video recording

1. Introduction

"ROV" (**Figure 1**) stands for remotely operated vehicle; ROVs are unoccupied, highly maneuverable underwater robots that can be used to explore ocean depths while being operated by someone at the water surface [1].

In a ROV, the connection between the vehicle and the surface is ensured by an umbilical cable that allows bi-directional communication, as well as energy supply to the vehicle. The use of this equipment in Underwater Technical Inspections, allows to reach greater depths and for a longer period than would be achieved using divers. In addition, it is possible to operate in contaminated waters that pose a risk to human life [2].

The vehicle is operated by the pilot from a command and control unit. This command includes two joysticks to control the depth and direction of the ROV, as well as commands to guide the video cameras (rotation and tilt), adjust the intensity of the lighting, control the articulated arm, and select the autopilot in direction or depth [2].

Figure 1.
Equipment used SIBIU PRO (NIDO ROBOTICS): Maximum depth of 300 meters.

The co-pilot assists in the navigation maneuver, as he is responsible for observing, analyzing and interpreting the sonar images and the acoustic positioning, giving indications to the pilot where to go [2].

The video signal is digitally recorded on magnetic tape, integrating information about the depth, the azimuth, the number of turns that the vehicle has accumulated on its own axis, as well as the date and time of the dive [2].

Most ROVs are equipped with at least a still camera, video camera, and lights, meaning that they can transmit images and video back to the ship. Additional equipment, such as a manipulator or cutting arm, water samplers, and instruments that measure parameters like water clarity and temperature, may also be added to vehicles to allow for sample collection [1].

First developed for industrial purposes, such as internal and external inspections of underwater pipelines and the structural testing of offshore platforms, ROVs are now used for many applications, many of them scientific. They have proven extremely valuable in ocean exploration and are also used for educational programs at aquaria and to link to scientific expeditions live via the Internet [1].

ROVs range in size from that of a small computer to as large as a small truck. Larger ROVs are very heavy and need other equipment such as a winch to put them over the side of a ship and into the water [1].

While using ROVs eliminates the "human presence" in the water, in most cases, ROV operations are simpler and safer to conduct than any type of occupied-submersible or diving operation because operators can stay safe (and dry!) on ship decks. ROVs allow us to investigate areas that are too deep for humans to safely dive themselves, and ROVs can stay underwater much longer than a human diver, expanding the time available for exploration [1].

2. Underwater technical inspections in Canary Islands

The underwater environment can be particularly harsh on structures, posing unique challenges to inspectors who must evaluate scour, material conditions or construction [3].

The technical inspection was carried out by Pharos Company using an underwater drone "ROV" (Remote Operated Vehicle), an unmanned underwater robot connected to a surface control unit by means of an umbilical cable.

The ROV used in this inspection is the "SIBIU PRO" developed by the company NIDO ROBOTICS, equipped with an HD camera, with 300 m of umbilical cable and four lights of 1,500 lumens. This ROV allows diving to a maximum depth of 300 m.

The following reports pretend to illustrate the reliability of Underwater Technical Inspections developed by similar companies around the world resorting to ROVs based on the study case of Canary Islands.

2.1 Canary Islands

The Canary Islands, also known informally as the Canaries, are a Spanish archipelago and the southernmost autonomous community of Spain located in the Atlantic Ocean, in a region known as Macaronesia, 100 km (62 miles) west of Morocco at the closest point (**Figure 2**). It is one of eight regions with special consideration of historical nationality as recognized by the Spanish government [4, 5].

The eight main islands are (from largest to smallest in area) Tenerife, Fuerteventura, Gran Canaria, Lanzarote, La Palma, La Gomera, El Hierro and La Graciosa (**Figure 3**). The archipelago includes many smaller islands and islets: Alegranza, Isla de Lobos, Montaña Clara, Roque del Oeste, and Roque del Este. It also includes a series of adjacent rocks (those of Salmor, Fasnia, Bonanza, Garachico and Anaga). In ancient times, the island chain was often referred to as "the Fortunate Isles" [6].

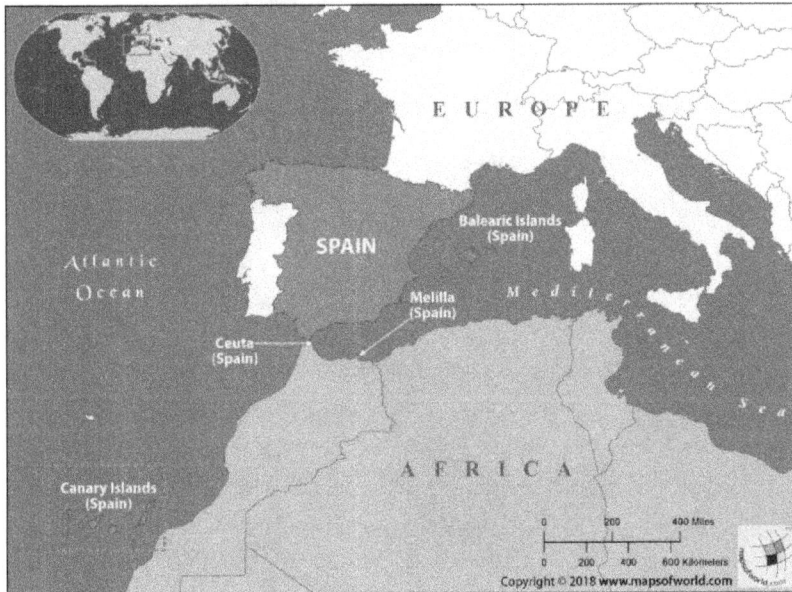

Figure 2.
Spain (source: www.mapsofworld.com).

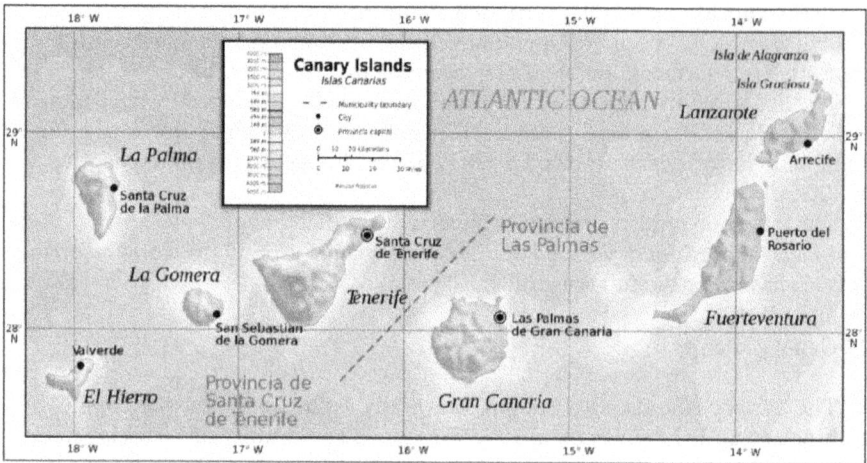

Figure 3.
The Canary Islands (source: www.zonu.com).

2.2 Inspection of interior docks (reinforced concrete quay blocks)

Two dives were carried out to control the execution of the interior docks expansion work (second phase), in the Las Palmas Port (**Figures 4–10**).

During the inspections, the following elements were visually controlled:

- Foundation of the quay blocks (bench, foundation and guard blocks);

- Condition of the concrete block wall as it closes to the RO-RO ramp;

- General condition of the vertical facing of the quay blocks;

- Completion of the lips of the docking superstructure;

- Location of wreck inside the Nelson Mandela dock.

Figure 4.
Interior docks, Port of Las Palmas.

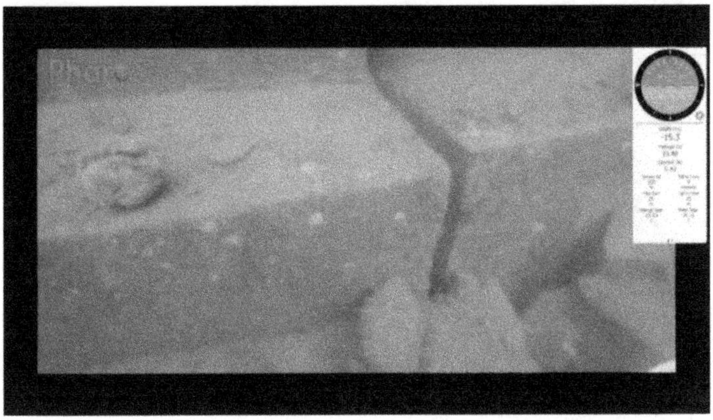

Figure 5.
Quay blocks joint detail. Foundation footing.

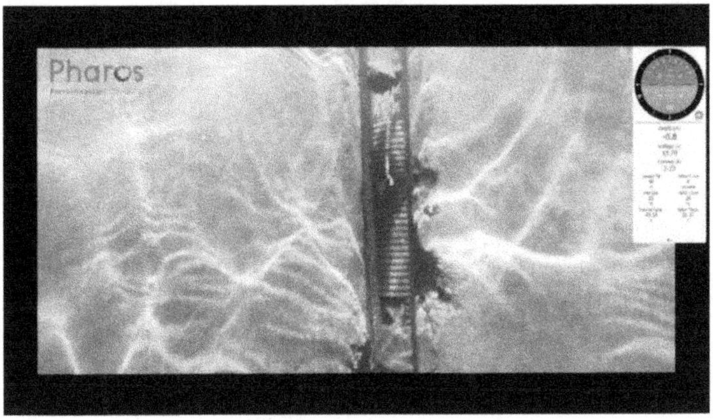

Figure 6.
PVC pipe detail, quay blocks joint.

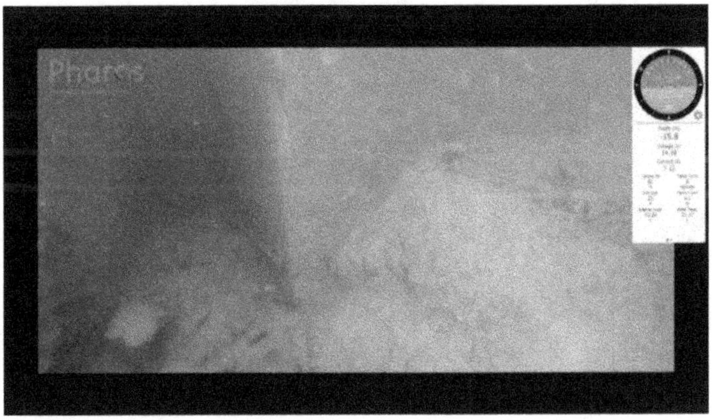

Figure 7.
Detail of guard concrete executed in foundation.

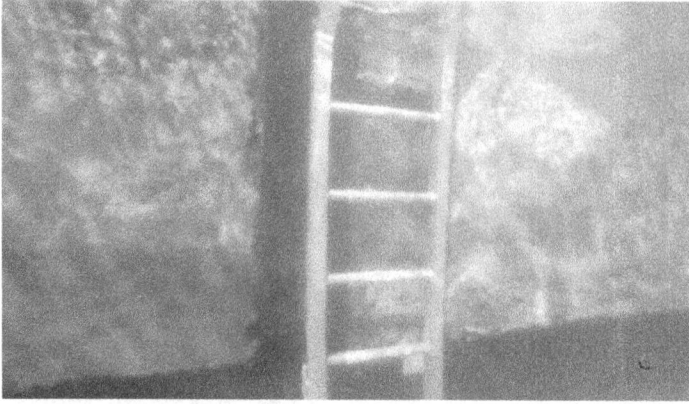

Figure 8.
Lip finish of superstructure in submerged area.

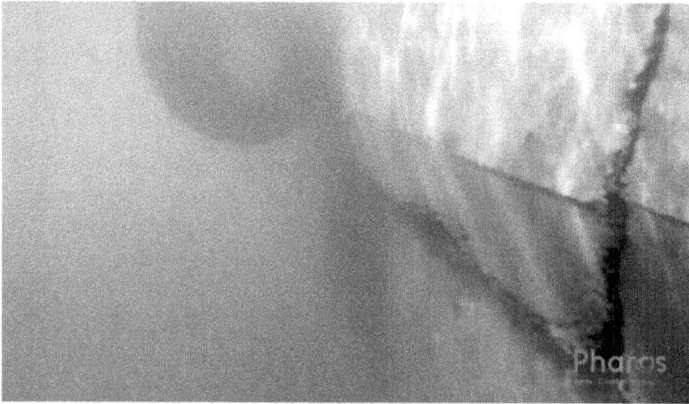

Figure 9.
Defense detail and lip concreting joints executed with continuous trolley.

Figure 10.
Detail of the wreck found.

2.3 Ro-Ro ramp inspection (bulk concrete blocks)

Two dives were carried out with the ROV to visually check the initial and final state of the repair of the berthing ramp of the passenger ships of the Shipping companies that operate in the Port of Las Palmas (**Figures 11–15**).

The elements to check were:

- Foundation of the bulk concrete blocks;

- General condition of the vertical wall, consisting of bulk concrete blocks;

- State of the finish of the ramp in its submerged part.

Figure 11.
Ro-Ro ramp, Port of Las Palmas.

Figure 12.
Detail of joint between concrete quay blocks in vertical face.

Figure 13.
Detail of the foundation of the ramp.

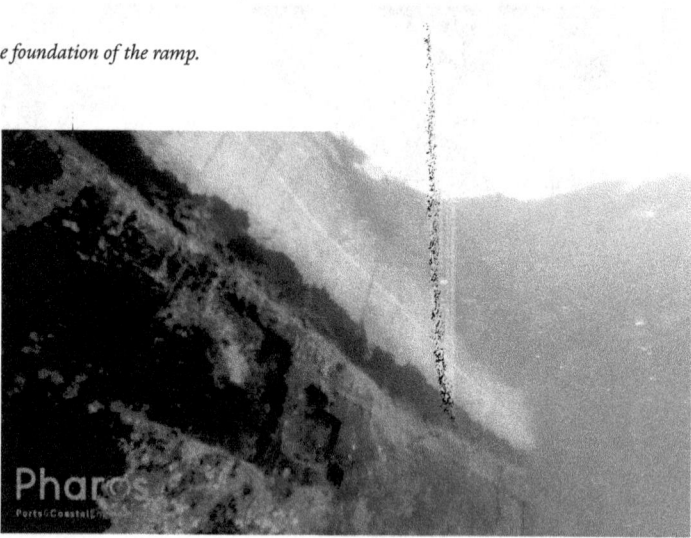

Figure 14.
Ramp repair area.

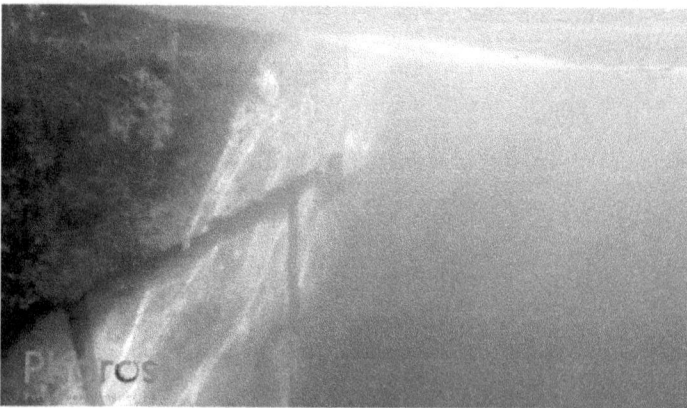

Figure 15.
Vertical face of the ramp.

2.4 Inspection of shelter dikes (reinforced concrete quay blocks)

2.4.1 Port of Las Palmas, Gran Canaria

An immersion was carried out with the ROV to visually check the state of the outer dock of the Port of Las Palmas, in the hammer area (**Figures 16–18**).
The items inspected were:

- Condition of the exterior and interior joints between quay blocks that make up the hammer;

- Condition of the foundation of the reinforced concrete quay blocks that make up the hammer;

- Condition of the concrete guard blocks in the exterior area.

Figure 16.
Shelter dike, Port of Las Palmas.

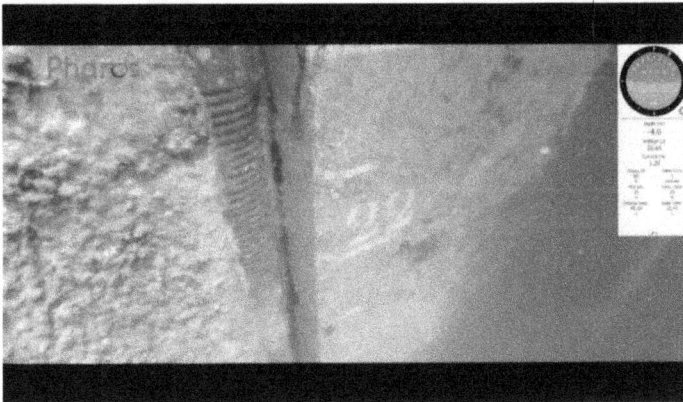

Figure 17.
Detail of depth markers on the facing of concrete quay blocks and joint.

Figure 18.
Guard concrete blocks on the foundation bank of the external dike.

2.4.2 Port of Arrecife, Lanzarote

A visual inspection of the submarine emissary of Arrecife was carried out by ROV (**Figures 19–22**).
The items inspected were:

- General inspection of the entire layout of the submarine emissary;

- Checking the state of the joints of the different sections of the pipe;

- Search for possible leaks in the pipe section;

- State of the concrete weights;

- Inspection of the state of the diffusers.

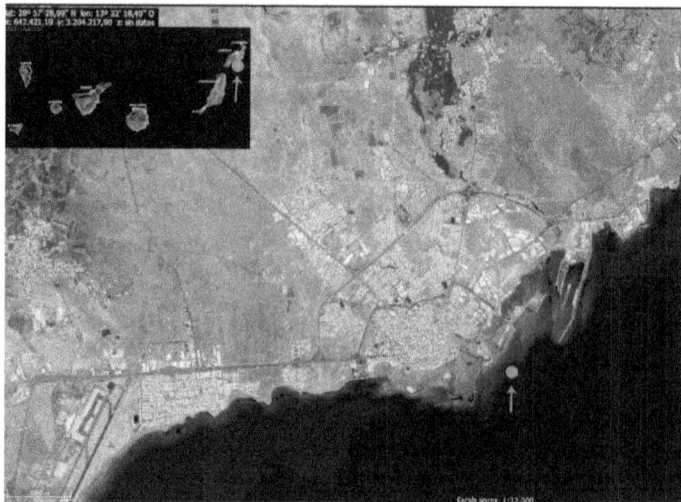

Figure 19.
Shelter dike, Port of Arrecife.

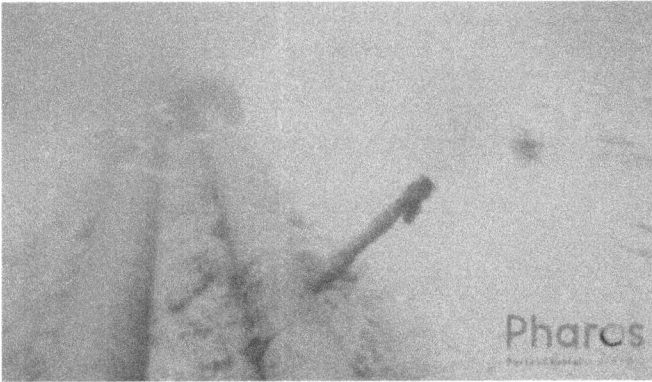

Figure 20.
Detail of the diffuser system of the submarine emissary pipeline.

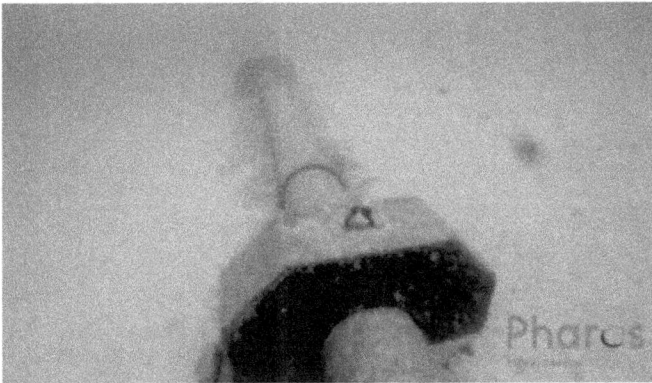

Figure 21.
Details of weights over the submarine emissary pipeline.

Figure 22.
Detail of the concrete weights in a pipeline section near the coast.

2.5 Foundation slab inspection - Duke of Alba

Three visual inspections were carried out through ROVs during the construction and completion phases of the expansion work for the cruise berth in the Port of Naos (Arrecife) (**Figures 23–26**).

The Duke of Alba consists of a reinforced concrete slab and a superstructure on reinforced concrete piles with lost casing.

The elements to check were:

- Starting state of the piles on the foundation slab;

- Condition of the surface and perimeter foundation slab;

- Condition of the foundation bench;

- Estimation of the height of the executed foundation slab.

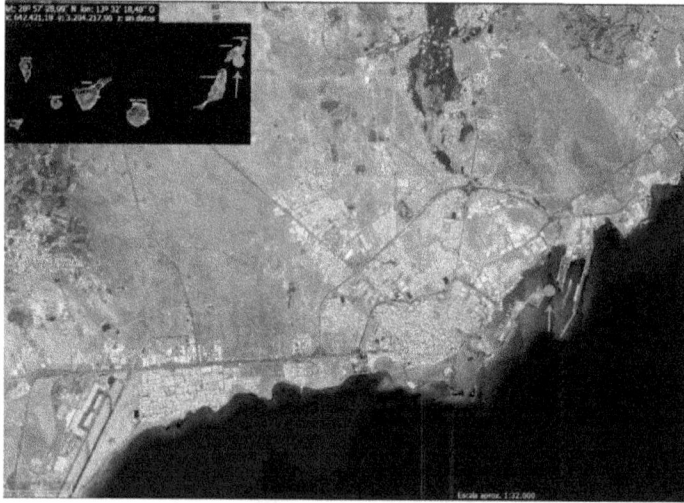

Figure 23.
Foundation slab – Duke of Alba, Port of Arrecife.

Figure 24.
Duke of Alba's reinforced concrete foundation slab (Port of Naos).

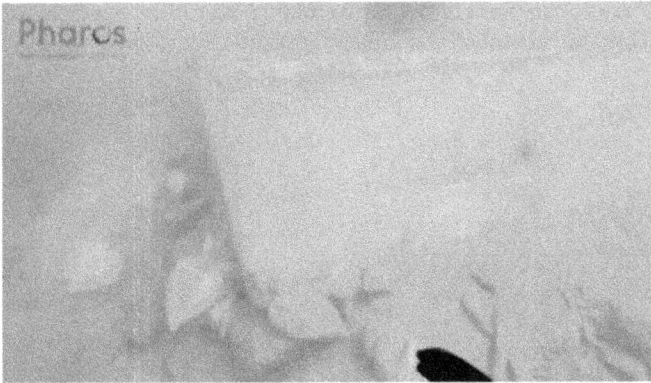

Figure 25.
Corner detail of the Duke of Alba's foundation slab (Port of Naos).

Figure 26.
Detail of upper surface and pile in the foundation of the Duke of Alba (Port of Naos).

3. Conclusions

In civil engineering, supervision, control, measurement and assessment of all phases of the work are essential: from project conception, planning, execution, including preservation and maintenance of infrastructures. This integral management model makes it possible to optimize resources and ensure that quality standards are achieved for works in progress and in service.

One of the handicaps that maritime work has had historically is the inherent difficulty of being partially or totally submerged in the aquatic environment. Underwater robotics and the reduction of the costs by using ROV equipment can constitute a turning point in the way of conceiving the management of works, quality and maintenance of the different coastal and port infrastructures.

The State Ports Administration - Spain (in Spanish: *Administración de Puertos del Estado*) and specifically the competent Port Authorities in the Canary Islands, have been promoting the use of the ROV as a new alternative for the management, supervision and maintenance of its infrastructures for a few years. The ROV is a highly reliable and safe, automatable and configurable technology that dramatically reduces costs and risks in underwater inspection operations.

The ROV is an equipment of the future and with a future, which is already an essential part in the present of maritime works and which, undoubtedly, is here to stay.

Author details

Sérgio António Neves Lousada[1,2,3,4*], Rafael Freitas Camacho[1,5] and Josué Suárez Palacios[6]

1 Department of Civil Engineering and Geology (DECG), Faculty of Exact Sciences and Engineering (FCEE), University of Madeira (UMa), Funchal, Portugal

2 VALORIZA - Research Centre for Endogenous Resource Valorization, Portalegre, Portugal

3 Institute of Research on Territorial Governance and Inter-Organizational Cooperation, Dąbrowa Górnicza, Poland

4 CITUR - Madeira - Centre for Tourism Research, Development and Innovation, Madeira, Portugal

5 IHM - Investimentos Habitacionais da Madeira, EPERAM, Portugal

6 Pharos Port and Coastal Engineering, Las Palmas de Gran Canaria, Spain

*Address all correspondence to: slousada@staff.uma.pt

IntechOpen

References

[1] NOAA, "What is an ROV?," NOAA, 2020. [Online]. Available: https://oceanexplorer.noaa.gov/facts/rov.html. [Acessed on 6 June 2020].

[2] Instituto Hidrografico, "ROV (Remotely Operated Vehicle)," Instituto Hidrografico, 2020. [Online]. Available: https://www.hidrografico.pt/info/19. [Acessed on June 2020].

[3] KCI, "Underwater Inspection and Commercial Diving," KCI, 2020. [Online]. Available: https://www.kci.com/services/transportation/structural-inspection/underwater-inspection-marine-engineering/. [Acessed on 6 June 2020].

[4] Gobierno de Canarias, "Reforma del Estatuto de Autonomía de Canarias," Gobierno de Canarias, 15 May 2006. [Online]. Available: https://web.archive.org/web/20060515081511/http://www.gobcan.es/reformaestatuto/index.aspx. [Acessed on 6 June 2020].

[5] Gobierno de Canarias, "Canarias en la España contemporánea: La formación de una nacionalidad histórica.," Gobierno de Canarias, 7 July 2016. [Online]. Available: http://coloquioscanariasamerica.casadecolon.com/index.php/CHCA/article/viewFile/7612/6583. [Acessed on 6 June 2020].

[6] T. Benjamin, The Atlantic World: Europeans, Africans, Indians and Their Shared HIstory, 1400-1900, Cambridge University Press., 2009, p. 107.

www.ingramcontent.com/pod-product-compliance
Lightning Source LLC
Chambersburg PA
CBHW081241190326
41458CB00016B/5868